Faceted Search

Synthesis Lectures on Information Concepts, Retrieval, and Services

Editor
Gary Marchionini, *University of North Carolina, Chapel Hill*

Faceted Search
Daniel Tunkelang
www.morganclaypool.com

ISBN: 9781598299991 paperback

ISBN: 9781608450008 ebook

DOI: 10.2200/S00190ED1V01Y200904ICR005

A Publication in the Morgan & Claypool Publishers series

SYNTHESIS LECTURES ON INFORMATION CONCEPTS, RETRIEVAL, AND SERVICES

Lecture #5

Series Editor: Gary Marchionini, University of North Carolina

Series ISSN
ISSN 1947-945X   print
ISSN 1947-9468   electronic

Faceted Search

Daniel Tunkelang
Endeca

SYNTHESIS LECTURES ON INFORMATION CONCEPTS, RETRIEVAL, AND SERVICES # 5

MORGAN & CLAYPOOL PUBLISHERS

ABSTRACT

We live in an information age that requires us, more than ever, to represent, access, and use information. Over the last several decades, we have developed a modern science and technology for information retrieval, relentlessly pursuing the vision of a "memex" that Vannevar Bush proposed in his seminal article, "As We May Think."

Faceted search plays a key role in this program. Faceted search addresses weaknesses of conventional search approaches and has emerged as a foundation for interactive information retrieval. User studies demonstrate that faceted search provides more effective information-seeking support to users than best-first search. Indeed, faceted search has become increasingly prevalent in online information access systems, particularly for e-commerce and site search.

In this lecture, we explore the history, theory, and practice of faceted search. Although we cannot hope to be exhaustive, our aim is to provide sufficient depth and breadth to offer a useful resource to both researchers and practitioners. Because faceted search is an area of interest to computer scientists, information scientists, interface designers, and usability researchers, we do not assume that the reader is a specialist in any of these fields. Rather, we offer a self-contained treatment of the topic, with an extensive bibliography for those who would like to pursue particular aspects in more depth.

KEYWORDS

faceted search, exploratory search, information seeking, human–computer information retrieval

Preface

We live in an information age, a world where, more than ever, we need to understand how to represent, access, and use information. Over the last several decades, we have witnessed the development of a modern science and technology for information access. We've come a long way from the vision of a "memex" that Vannevar Bush proposed in his seminal article, "As We May Think" [1]: a mechanical device that would allow someone to access a large, self-contained research library.

Many people may feel that we have already achieved that end goal, albeit by different means. After all, we can now enter the name of a person or company into Google or some other web search engine and, in most cases, be instantly directed to the associated web page. Thanks to tools like the collectively edited Wikipedia, we may achieve similar success with more general queries, at least for topics of broad enough interest to have inspired Wikipedia entries.

Looking beyond these use cases, however, we see that we are only in the early days of implementing the memex vision (Figure 1). Modern search engines adequately address the problem of what library scientists have historically called known-item search: we know what we are looking for and are certain it exists in the collection we are searching [2].

In contrast, we have not developed comparably mature tools for exploratory search—that is, information seeking where users do not have a known target document and may not even have a well-established information need [4]. Only in the last few years have we seen an emerging program of human–computer information retrieval (HCIR) that brings interactive techniques—many inspired by pre-Internet research in library science—to bear on more sophisticated information-seeking tasks [5].

Facets, a way of classifying information, play a key role in this program. Faceted classification addresses the weakness of earlier knowledge representations—namely, the rigidity of taxonomical schemes and the chaos of unstructured indexes. Developed by library scientists, faceted classification offers an approach to knowledge representation that is both faithful to its richness and practical for real-world use.

Faceted classification, however, only addresses the problem of representing information. We still need a means to access and use that information. That means is faceted search.

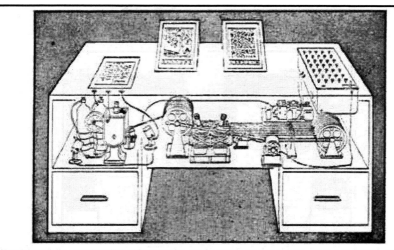

Memex in the form of a desk would instantly bring files and material on any subject to the operator's fingertips. Slanting translucent viewing screens magnify supermicrofilm filed by code numbers. At left is a mechanism which automatically photographs longhand notes, pictures and letters, then files them in the desk for future reference (*LIFE* 19(11), p. 123).

FIGURE P.1: Vannevar Bush's theoretical memex machine [3].

Researchers such as Marti Hearst have led the way with user studies that demonstrate how faceted search provides more effective information-seeking support to users than conventional best-first search—even though users are more familiar with the latter [6].

Those dubious of the value of faceted search interfaces raise the specter of what Joshua Porter and his colleagues at User Interface Engineering call the "three-click rule,"—that is, the web design rule of thumb that no piece of content should take more than three clicks to access. Luckily, that bit of folk wisdom does not hold up to empirical study [7]. When tested in a user study, it was found that there is no correlation between the number of times users clicked and their success in finding the content they sought, and that the number of clicks is not what is important to users, but only whether or not they are successful at finding what they are seeking (Figure 2).

In the lecture that follows, we will explore the history, theory, and practice of faceted search. Although faceted search has become increasingly prevalent in online information access systems, this text is, to our knowledge, the first comprehensive treatment of the subject. Although we cannot hope to be exhaustive, our aim is to provide sufficient depth and breadth to offer a useful resource to both researchers and practitioners.

Because faceted search is an area of interest to computer scientists, information scientists, interface designers, and usability researchers, we do not assume that the reader is a specialist in any of

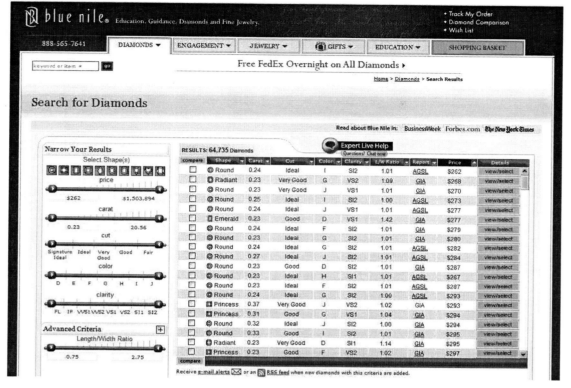

FIGURE P.2: Results for "diamonds" at www.bluenile.com.

these fields. Rather, we offer a self-contained treatment of the topic, with an extensive bibliography for those who would like to pursue particular aspects in more depth.

The book consists of three parts. The first part presents the key concepts leading the reader on a historical path from Aristotle's classical ideas of knowledge representation to a modern-day definition of faceted search. The second part describes key work on faceted search in academia and industry. The third part addresses some of the practical challenges that confront the developers of faceted search applications.

Each chapter ends with take-aways that summarize the chapter's key points. Impatient readers may skip to these, but I hope you will find the journey as valuable as the destination.

Acknowledgments

I thank Gary Marchionini, Diane Cerra, and all of the kind people at Morgan Claypool for inviting me to write this lecture. I am honored to participate in the Synthesis Lectures on Information Concepts, Retrieval, and Services, which represent a fantastic collection of work by the leading researchers in information science and retrieval.

I thank Candy Schwartz, Co-Editor-in-Chief of *Library & Information Science Research*, for reviewing an early draft of this lecture. I am also indebted to a veritable army of volunteer reviewers: Omar Alonso, Pete Bell, Amitava Biswas, Blade Kotelly, Sol Lederman, Milan Merhar, Jennifer Novosad, XiaoGuang Qi, Brett Randall, Dusan Rnic, and Joshua Young.

I thank my colleagues and management at Endeca for allowing me to take time away from my day job to try my hand as a writer. I also thank Endeca's customers, whose applications made many of the figures in this book possible.

Finally, I am grateful to my wife Kristin and my daughter Lily for supporting me and putting up with me in these manic months.

Contents

PART I

Key Concepts

The first part of this book introduces the key concepts necessary to provide a foundation for faceted search. Readers who are already familiar with the basic ideas of knowledge representation and information retrieval may find this part of the book unnecessary; conversely, readers unfamiliar with these ideas will hopefully learn enough in this brief introduction to make sense of and derive utility from the remainder of the lecture.

CHAPTER 1

Introduction: What Are Facets?

"Science is organized knowledge."

—Immanuel Kant

Before diving into the details of faceted search, we will establish what facets are in the first place. Specifically, we will discuss classification in general, and faceted classification in particular.

In this introductory chapter, we very briefly offer a history of knowledge representation that spans almost two and a half millennia. Our purpose is to place faceted classification, the knowledge representation at the heart of faceted search, in the context of this history.

1.1   CLASSIFICATION: ARISTOTLE'S TREE

Aristotle may have been the last person to know everything there was to be known in his own time [8]. He created the first comprehensive system of philosophy, encompassing morality, aesthetics, logic, the physical sciences, and even politics.

Of interest to us, however, is that Aristotle was among the first to establish a framework for this collective knowledge of the human race. If, as Whitehead asserts [9], all philosophy is "a series of footnotes to Plato," then all theory and practice of knowledge representation are surely a series of footnotes to Aristotle.

Aristotle's system of classifying living things divided organisms into two groups, plants and animals; further dividing animals into those "with blood" and "without blood"; those with blood into live-bearing vs. egg-bearing; and so forth (Figure 1.1). Aristotle's *Historia Animalium* [11], *De Partibus Animalium* [12], and *De Generatione Animalium* [13] framed the science of zoology for the next two millennia.

Aristotle was the first taxonomist—and his role as such would be familiar to those who work as taxonomists today, organizing knowledge into hierarchies. The word *taxonomy* comes from the Greek τάξις (*taxis*) meaning order or arrangement and νόμος (*nomos*) meaning law or science. It originally referred to the classification of living things, as per the Aristotelian taxonomy just described and the 18th century Linnaean taxonomy *Systema Naturae*, named for its author Carolus Linnaeus [14].

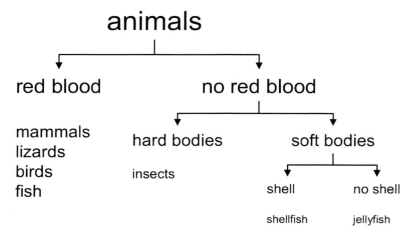

FIGURE 1.1: A subset of Aristotle's classification system [10].

Today, however, the word *taxonomy* refers more generally to any hierarchical classification scheme, as well as to the principles underlying such schemes (Figure 1.2).

In modern use, taxonomy is any organization of things or abstractions into a hierarchy, or tree structure. In keeping with the tree as a metaphor (albeit an upside-down tree), there is a root node at the top, leaf nodes at the bottom, and branches connecting each nonleaf parent node to its children (Figure 1.3). A parent may have many children, but each nonroot node has exactly one parent.

A representation in which a node may have multiple parents—and thus multiple disjoint paths leading to it from the root—is called a *polyhierarchy* [16]. Polyhierarchy introduces additional expressivity (e.g., a spork can be both a spoon and a fork) but at the expense of considerable complexity. We will therefore limit taxonomies in the context of this lecture to those displaying strict hierarchies.

The root node in a taxonomy represents a catch-all classification that describes the entire collection of objects. For example, in the Aristotelian taxonomy, the root node corresponds to the set of all living things. Its children represent the top-level divisions of the collection; their children, subdivisions of those; and so forth, until the leaf nodes represent the objects themselves.

The key property of a taxonomy is that, for every object or set of objects that corresponds to a node, there is precisely one unique path to it from the root node. Thus, a taxonomy imposes a strict logical ordering on the knowledge that it represents.

The modern analogs of the Aristotelian and Linnaean taxonomies include the Dewey Decimal Classification [17] for libraries, as well as web directories like the Yahoo! Directory [18] and the Open Directory Project [19], both of which aspire to catalog large subsets of the vast collection of sites available on the World Wide Web. It is worth noting that in all three cases, the object of arrangement is to organize items (books on shelves, web sites in a list) rather than to organize abstract concepts.

Animals are either,

Sanguineous, that is, such as have Blood, which breathe either by
Lungs, having either
Two Ventricles in their Heart, and those either
Viviparous.
Aquatick, as the Whale-kind,
Terrestrial, as Quadrupeds.
Oviparous, as Birds.
But one Ventricle in the Heart, as Frogs, Tortoises, and Serpents.
Gills, as all sanguineous Fishes, except the Whale-kind.
Exanguineous, or without Blood, which may be divided into
Greater; and those either
Naked.
Terrestrial, as naked Snails,
Aquatick, as the Poulp, Cuttle-fish, &c.
Covered with a Tegument, either
Crustaceous, as Lobsters and Crab-fish.
Testaceous, either
Univalve, as Limpets,
Bivalve, as Oisters, Muscles, Cockles, &c.
Turbinate, as Periwinkles, Snails, &c.
Lesser, as Insects of all Sorts.

Viviparous hairy Animals, or Quadrupeds, are either
Hoof'd, which are either
Whole-footed or hoof'd, as the Horse and Ass:
Cloven-footed, having the Hoof divided into
Two principal Parts call'd Biscula, either
Such as chew not the Cud, as Swine.
Ruminant, or such as chew the Cud, divided into
Such as have perpetual and hollow Horns:
Beef-kind,
Sheep-kind,
Goat-kind.
Such as have solid, branched, and deciduous Horns, as the Deer kind.
Four Parts, or Quadriscula, as the Rhinoceros and Hippopotamus.
Claw'd, or digitate, having the Foot divided into
Two Parts or Toes, having two Nails, as the Camel-kind.
Many Toes or Claws, either
Undivided, as the Elephant
Divided, which have either
Broad Nails and an humane Shape, as Apes.
Narrower and more pointed Nails, which in respect of their Teeth, are divided into such as have
Many Fore-teeth or Cutters in each Jaw:
The greater, which have
A shorter Snout and rounder Head, as the Cat-kind.
A longer Snout and Head, as the Dog-kind.
The lesser, the Vermin or Weazel kind.
Only two large and remarkable Fore-teeth, all which are Phytivorous and are call'd the Hare-kind.

FIGURE 1.2: Ephraim Chambers's *Cyclopaedia* (1728) [15].

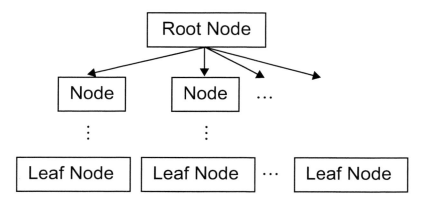

FIGURE 1.3: A generic taxonomy.

The strict ordering of a taxonomy can be too rigid. As the world's collection of knowledge has grown, taxonomies have struggled to keep pace. In particular, the requirement that every node in a taxonomy have a unique path from the root node is a harsh constraint, imposing Solomonic choices on those who design taxonomies. For example, is "European History" a child of "Europe" or of "History"? In general, compound concepts pose a challenge for the hierarchical organization of taxonomies.

As we briefly mentioned, polyhierarchies offer a work-around, but the introduction of polyhierarchy creates more problems than it solves, particularly for those with the thankless job of maintaining them. It is difficult enough to maintain a standard taxonomy, and the difficulty becomes that much greater when moving a node does not simply move its subtree with it.

In the next section, we will see how faceted classification provides a more elegant solution to the challenge of compound concepts.

1.2   FACETS: RANGANATHAN'S COLONS

Aristotle's approach to classification held sway for two millennia, ultimately inspiring Melvil Dewey to develop the Dewey Decimal Classification (DDC) system, a taxonomy still used by many libraries today. As of 2003, there were 110,000 leaf nodes in the DDC taxonomy [20].

A node in the DDC taxonomy is a sequence of decimal digits, with (in most cases) each successive digit corresponding to following a branch in the tree. For example, consider the following path to the "Cats" node:

<u>6</u>00	Technology
<u>63</u>0	Agriculture and related technologies

<u>636</u>	Animal husbandry
<u>636.8</u>	Cats

Books about cats are classified under Technology?! Actually, it turns out that cats have (at least) two homes in the DDC: 636.8 for cats in the context of animal husbandry, and 599.75 for cats in the context of zoology (500 is Science). Where do books about cat behavior go? That depends on the cataloguer's world view.

Here we can see the challenge inherent in using a single hierarchical taxonomy to represent a diverse collection of knowledge. Each branch in the hierarchy requires us to make an indelible choice, and we are particularly subject to the arbitrary divisions made earliest in the classification process.

The person who most clearly saw this problem was Shiyali Ramamrita Ranganathan. An Indian mathematician and librarian, S. R. Ranganathan lived from 1892 to 1972. In 1957, he won the Padma Shri, one of India's highest civilian awards, for his contributions to library science [21].

In his seminal text, *Philosophy of Library Classification* [22], Ranganathan discusses *hospitality of notation*, a term he attributes to Charles Ammi Cutter [23]. He takes the perspective of a librarian who has to maintain a taxonomy as new subjects appear.

Ranganathan writes:

A system of notation that does not admit of interpolation and extrapolation may be said to be inhospitable or rigid. One that admits of it may be said to be hospitable or elastic.

Ranganathan continues to explain that the DDC's hospitality of notation resides at the right end of the notation—that is, the only means to establish a new classification is to insert it as a child node of an existing node in the hierarchy. In the DDC, this insertion corresponds to specifying the right-most digit of the node: either replacing a zero (0) by a nonzero digit or adding a new digit to the right of the decimal point.

What Ranganathan wanted, however, was a notation that provided "hospitality at many points"—not just at the right end of the notation. In particular, he wanted a notation that could accommodate a general class of compound subjects—something not possible in a hierarchical taxonomy because of the aforementioned uniqueness of paths.

To this end, Ranganathan introduced the colon classification scheme in 1933 [24]. Attributing his inspiration to the extensibility of a Meccano set, a child's toy set he had seen in a London department store, Ranganathan developed the first library classification scheme based on facet analysis.

And thus we arrive at the core concept of this book: facets. Ranganathan did not even call them facets; he used the term *isolates* to describe composable elements ideally suited to constructing compound subjects, much as the Meccano set offered the basic components that children could combine to build complex mechanical objects, such as model locomotives. For clarity, however, we will refer to his isolates as facets.

Ranganathan decomposed the world of knowledge into five fundamental categories—personality, matter, energy, space, and time (known as PMEST)—which form the basis for facet analysis. His notation expresses a compound subject as a sequence of symbols (letters and numbers) separated by colons—(and, as the classification scheme developed, other punctuation)—hence, the name *colon classification*. Each individual facet is a hierarchy formed using various characteristics of division (e.g., by subject, by ownership, by community served, and so on), and the foci in the hierarchies can be combined in a prespecified order to form compounds.

In *Classification, Coding, and Machinery for Search* [25], Ranganathan gives an example of such a compound subject: L2153:4725:63129:B28 represents the statistical study of the treatment of cancer of the soft palate by radium.

We can break down this compound subject into its four constituent hierarchical facets:

- Medicine (L) → Digestive system (L2) → Mouth (L21) → Palate (L215) → Soft palate (L2153)
- Disease (4) → Structural disease (47) → Tumor (472) → Cancer (4725)
- Treatment (6) → Treatment by chemical substances (63) → Treatment by a chemical element (631) → Treatment by a group 2 chemical element (6312) → Treatment by radium (63129)
- Mathematical study (B) → Algebraical study (B2) → Statistical study (B28)

As we can see from this example, the colon classification system is extensible beyond the normal application of the DDC or any other classification scheme restricted to a single hierarchical taxonomy. The decomposition of subjects into component facets makes it possible to extend each facet independently.

This extensibility, or "hospitality at many points," is Ranganathan's gift to library science and has inspired modern approaches to thesaurus construction [26]. More importantly to our purposes, faceted classification makes it possible to build faceted search systems that support information seeking in a way reminiscent of librarians' reference interviews.

1.3   ONTOLOGIES

Although the focus of this lecture is faceted search, we would be remiss, however, not to touch on representations that extend beyond faceted classification and thus suggest information access techniques that push the boundaries of faceted search. Specifically, some information scientists, encountering the expressive limits of faceted classification systems, have instead turned to ontologies.

Like taxonomy, the word *ontology* comes from the Greek ὄντος (*ontos*) meaning of being and λογία (*logos*) meaning science, study, theory. In philosophy, ontology is the study of the nature of being.

In information science, the word *ontology* has taken on a more practical meaning: a representation of a set of concepts and the relationships among those concepts [27]. A taxonomy is a special case of an ontology, where the only relationships are "is–a" relationships connecting a child node to its parent. As we have seen in the previous section, a faceted classification system comprises a collection of facets, each of which is a taxonomy.

Ontologies, however, allow for other, more general relationship types. For example, we may have Homer is-author-of *The Iliad* or Paris is-capital-of France. In an ontology, a relationship is simply a tuple indicating the related objects and the relationship type.

In fact, an ontology can even include tuples that combine more than two objects. For instance, there may be a complex tuple representing an Academy Award, for example, Dustin Hoffman winning Best Actor for his 1988 performance as Raymond Babbitt in the movie *Rain Man*. In practice, however, these complex associations are factored into triples of the form (entity, relationship, entity).

In 2001, the World Wide Web Consortium created the Web Ontology Working Group that ultimately delivered the Web Ontology Language (OWL) in 2004 [28]. OWL has become a standard for the semantic web, a vision led by Tim Berners-Lee to extend the World Wide Web into a system that defines the semantics of information and services available on the web, with the goal of performing computation and deriving inferences [29]. This project is a work in progress, but the scope of its ambition makes it clear how important it is to choose an appropriate knowledge representation.

1.4   TAKE-AWAYS

- Knowledge representation is a problem whose study dates back two millennia to Aristotle and is still an active research area.
- Taxonomies are essential tools, but the constraint of each node having a unique path limits their expressivity and extensibility.
- Faceted classification decomposes compound subjects into foci in component facets, offering expressive power and flexibility through the independence of the facets.
- Ontologies and the semantic web offer the prospect of even richer possibilities for knowledge representation.

CHAPTER 2

Information Retrieval

"The seeking is the goal and the search is the answer."

—Anonymous

Having considered the characteristics and motivations of systems for representing information, we now turn to the problem of making that information accessible.

In this chapter, we briefly review the prevalent approaches to information retrieval. We first consider the dominant approaches for text search: set retrieval and ranked retrieval. We then consider directory-based navigation: an approach that, although not specific to text collections, has often been applied to them. In the following chapter, we turn to approaches that take advantage of faceted information models.

This short chapter makes no attempt to offer an exhaustive treatment of information retrieval—a subject that spans at least six decades, from Vannevar Bush's memex vision to the web search engines, such as those provided by Google, Yahoo!, and Microsoft, that are an indispensable part of our lives today. Moreover, we focus on the basic methods of document retrieval, ignoring areas like multimedia and social search.

If you are interested in learning more about this subject, I encourage you to read Amit Singhal's history of information retrieval—as well as the works he cites [30].

2.1   RELEVANCE

The core concern of information retrieval is to help users retrieve documents that are relevant to their information needs. The notion of relevance, however, is difficult to define. As William Goffman wrote in 1964 [31]:

Relevance is defined as a measure of information conveyed by a document relative to a query ... the relationship between the document and the query, though necessary, is not sufficient to determine relevance.

Stefano Mizzaro and Tefko Saracevic have written histories of relevance, documenting the evolution of this concept among library scientists and information retrieval researchers who have often disagreed about how to measure it [32, 33]. The former tend to take a cognitive, user-centered approach, whereas the latter take a benchmark-oriented approach. This philosophical difference leads to different evaluation approaches, library and information scientists favoring user studies, and information retrieval researchers favoring the use of test collections, particuarly those maintained by the Text REtrieval Conference (TREC) [34].

Unfortunately, the TREC approach for measuring relevance has not proven effective for interactive information retrieval systems, and user studies can be prohibitively expensive [35]. Hence, infromation retrieval researchers accept a definition of relevance as a measure entirely conveyed by a document relative to a query (Goffman notwithstanding). Library and information scientists adhere to a user-centered approach, typically measuring the effectiveness of information-seeking support systems through user studies at a task (or higher) level rather than at the query level.

2.2   SET RETRIEVAL

Although search is ubiquitous today, the earliest search engines—or, more formally, information retrieval systems—worked very differently from their modern counterparts. Unlike most modern search engines, the earliest information retrieval systems used a set retrieval model [36]. Set retrieval systems return results that are unordered document sets rather than ordered sequences—there is no concept of relevance ranking.

This model is also known as a Boolean retrieval model because set retrieval systems typically allow users to specify their query expressions using Boolean operations (AND, OR, NOT). Many such systems have extended the Boolean syntax to include additional operators to specify term order or the proximity of terms within a document. Some systems also allow users to specify where in a document a term occurred (e.g., in the title field vs. in the author field), a topic we will return to when we discuss parametric and faceted search.

Figure 2.1 shows an example of a Boolean search interface, the advanced search page for the US Patent Office database of issued patents. Users can specify queries in a rich language that includes standard Boolean operators and restrict matches to particular document fields.

Despite this flexibility, set retrieval suffers from a fundamental weakness. As users attempt to express their information needs as Boolean queries, they often find themselves facing a choice between high precision and high recall but are unable to simultaneously achieve adequate precision and recall. Indeed, the difficulty that users face in formulating Boolean queries may help explain there rarity in Web search queries, even though most web search engines support Boolean operators [38].

Let us define these terms and then explain the trade-off between them.

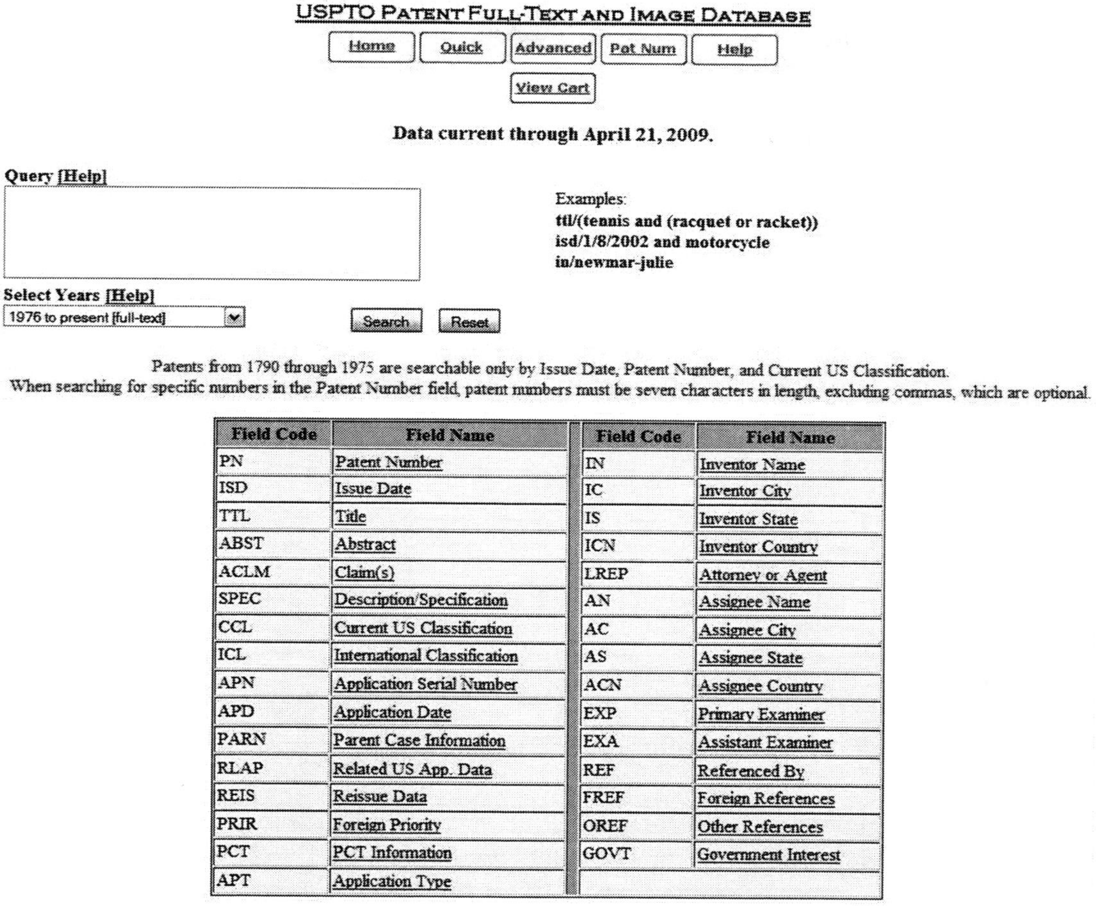

Field Code	Field Name	Field Code	Field Name
PN	Patent Number	IN	Inventor Name
ISD	Issue Date	IC	Inventor City
TTL	Title	IS	Inventor State
ABST	Abstract	ICN	Inventor Country
ACLM	Claim(s)	LREP	Attorney or Agent
SPEC	Description/Specification	AN	Assignee Name
CCL	Current US Classification	AC	Assignee City
ICL	International Classification	AS	Assignee State
APN	Application Serial Number	ACN	Assignee Country
APD	Application Date	EXP	Primary Examiner
PARN	Parent Case Information	EXA	Assistant Examiner
RLAP	Related US App. Data	REF	Referenced By
REIS	Reissue Data	FREF	Foreign References
PRIR	Foreign Priority	OREF	Other References
PCT	PCT Information	GOVT	Government Interest
APT	Application Type		

FIGURE 2.1: US Patent Office Boolean search interface [37].

2.3 PRECISION VS. RECALL

In the information retrieval literature, precision and recall are the two most prevalent measures for retrieval accuracy—often called *retrieval performance* in the literature, but we prefer the term *accuracy* because many people use "performance" to characterize speed or computational efficiency.

For a given query, *precision* is the fraction of retrieved documents that are relevant to the information need represented by the query. *Recall* is the fraction of all possible relevant documents that are retrieved. If we consider the judicial oath to "tell the truth, the whole truth, and nothing but the truth," recall measures the extent to which an information retrieval system tells the whole truth, and precision measures the extent to which it tells nothing but the truth.

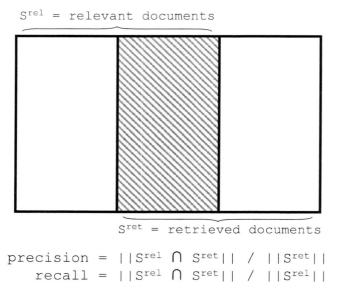

$$\text{precision} = ||S^{rel} \cap S^{ret}|| \,/\, ||S^{ret}||$$
$$\text{recall} = ||S^{rel} \cap S^{ret}|| \,/\, ||S^{rel}||$$

FIGURE 2.2: Precision and recall in the set retrieval model.

Figure 2.2 visualizes precision and recall in terms of the interaction of the sets of relevant and retrieved documents.

Ideally, a set retrieval system would achieve both high precision and high recall. As it turns out, this is difficult to do using a typical Boolean set retrieval interface. Let us use a concrete example to establish an intuition for this difficulty: imagine that we are searching for work on library science in a collection of scholarly literature.

We can achieve high precision by looking for works that contain the precise phrase "library science"—and can perhaps achieve close to 100% precision by only considering works that contain this phrase in their title. Stephen Robertson refers to such heuristics as "precision devices" [39].

Although a combination of precision devices may well achieve a sufficient degree of precision, it is likely to lead to poor recall. For example, none of Ranganathan's works cited in the previous chapter contain the phrase "library science" in their titles.

We might instead choose a strategy that emphasizes recall—for example, returning all works whose full text contains either "library" or "science"—or any variants of those words (e.g., librarian, scientist). We could increase recall further through thesaurus expansion (repository, collection, technology). Such an approach would undoubtedly achieve high recall, but at the expense of very low (e.g., less than 10%) precision.

Although the expansion described may seem so extreme as to be absurd, finding a suitable medium turns out to be difficult. Professional librarians are better than amateur information seekers at constructing complex Boolean queries to optimize the trade-off between precision and recall. Nonetheless, even they overestimate their power to make effective the use of Boolean set retrieval systems.

In fact, a 1994 study by the West Publishing Company (which provides the Westlaw information service to legal professionals) demonstrated that even experts were better off giving up Boolean retrieval in favor of simply entering words into a search box as a free-text query and relying on the search engine to return what it deemed to be the best matches [40]. It seems safe to assume that, in the years since the study, search engine developers have increased their algorithms' effectiveness more than expert information seekers have improved their query-formulating expertise.

2.4   RANKED RETRIEVAL

If set retrieval is too difficult for expert information seekers, what is the alternative? The alternative, popularized by modern search engines, is ranked retrieval, also known as relevance ranking.

In 1961, Calvin Mooers [41], whose claims to fame include his coining the term *information retrieval*, expressed his dissatisfaction with the Boolean set retrieval model that he helped establish [42]:

> It is a common fallacy, underwritten at this date by the investment of several million dollars in a variety of retrieval hardware, that the algebra of George Boole is the appropriate formalism for retrieval system design. This view is as widely and uncritically accepted as it is wrong.

In seeking an alternative to the Boolean retrieval model, information retrieval researchers took a completely different approach. Rather than requiring formal, structured queries, they pursued an approach based on unstructured, free-text queries, thus liberating users from the need to construct complex expressions. Rather than attempting to return a precise set of results to the user, they cast a wide net and rely on ranking to favor more relevant results.

Figure 2.3 shows an example of ranked retrieval at Rexa.info, a digital library and search engine covering computer science research literature. A query for "faceted search" returns an impressive 93,541 results; however, those results are for the query faceted OR search, which has high recall but low precision. The application developers mitigate this low precision through relevance ranking, and indeed the top-ranked results are far more relevant than those at the bottom of the list.

Free-text queries are, without a doubt, easier for users to create than formal Boolean expressions. But this liberation comes at a cost: the query no longer represents a well-defined filter on the document collection. Instead, the query becomes a target, and the binary notion of a document matching the query relaxes into a continuous similarity measure. An information retrieval system built on such an approach no longer filters documents but rather ranks them according to the degree to which they match the query.

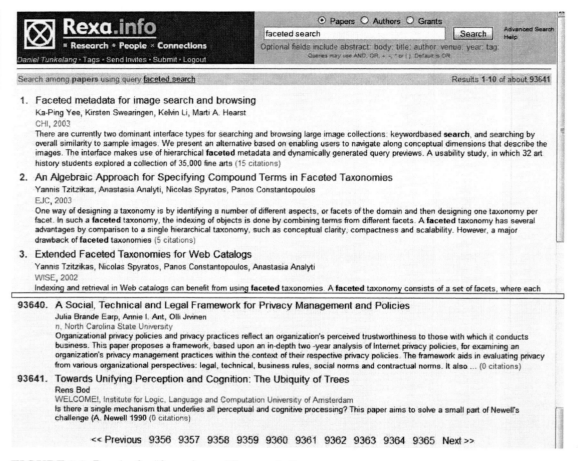

FIGURE 2.3: Results for "faceted search" at rexa.info.

The ranked retrieval model largely owes its existence and success to two of the greatest figures in modern information retrieval: Gerald Salton and Karen Spärck Jones.

Gerald Salton's contributions to information retrieval are so numerous that the highest honor for information retrieval research is the one that bears his name: the Gerald Salton Award. The particular contribution that interests us is the vector space model [43]. The vector space model represents each text document as a vector of words—or, more generally, terms that may include multiple-word phrases or truncated word stems. Each component of a document vector represents, at least in theory, the relative strength of is corresponding term, that is, the extent to which the document is about that term. Now all we need is a way to determine the values of these components.

That task fell to Karen Spärck Jones, another giant in the history of information retrieval. She proposed a statistical interpretation of term specificity that could then be applied to the vector space model [44]. This weighting scheme is known today as *term frequency-inverse document frequency*, or tf-idf. The hyphen is an unfortunate accident of notation, as the tf-idf multiplies the two factors and thus should be written tf*idf.

The first factor, term frequency, is the number of times a given term appears in that document divided by the number of terms in the document, to obtain a normalized score between zero and one. All else equal, a higher term frequency indicates that a term is more representative of the document's content.

All else, however, is not necessarily equal. The second factor, inverse document frequency, emphasizes rare terms over common ones. Terms that occur in fewer documents in a given corpus receive higher weight than terms that occur in more documents because scarcity implies specificity. The document frequency of a term is the fraction of the documents in the collection containing it, and the inverse document frequency is the reciprocal of the document frequency. Actually, despite the name, idf usually denotes the logarithm of this reciprocal.

Given the vector space model and tf-idf as a statistical interpretation of term specificity, we can treat a search query as a geometric problem of determining the distance of each document in a collection from the search query. More formally, we treat both the query and a document as vectors in a highly dimensional geometric space and compute the cosine of the angle between the two vectors. A smaller angle implies greater similarity, and the cosine of the angle provides a score that can be used for ranking.

There have been many developments in ranked retrieval since the early efforts of Salton and Spärck Jones. A major advance was the development of latent semantic analysis, a technique that applies singular-value decomposition to the term-document matrix to discover the hidden, or latent, topics that form the basis for a lower-dimensional vector space than the term vector space [45].

However, the most significant recent development in ranked retrieval models has had little to do with the query-dependent measures most associated with information retrieval research. Rather, the emergence of the World Wide Web as the document collection most targeted by search engines has led researchers to focus on hypertext collections in general and the web in particular.

The social construction of the web has led researchers to develop ranking approaches that emphasize query-independent measures of document authority, also known as document priors, to reflect their computation prior to a query. The two most notable authority measures for hypertext collection are Jon Kleinberg's HITS algorithm [46] and Larry Page and Sergey Brin's PageRank algorithm [47]—the latter having served as the initial foundation for Google's web search engine. Although authority is distinct from the concept of document relevance that motivated most of the work on ranked retrieval, it fits well into the ranked retrieval framework of sorting results by a utility score.

The success of ranked retrieval relative to set retrieval argues for the advantages of the former, but these advantages also come at a price. In a ranked retrieval model, we can no longer reason about which results match or do not match a query—we lose the bright line of the set retrieval model.

Although this loss may not seem important in an interface where users only look at top-ranked search results, we will see that this lack of a clear dividing line is highly problematic for faceted search. Faceted search is, at heart, a set-oriented retrieval method.

2.5   DIRECTORY NAVIGATION

In less than a decade, we have become accustomed to free-text search as the most familiar interface for information seeking. It is easy to forget that there are other interfaces that serve this purpose.

In the previous section, we discussed taxonomies, an approach to knowledge representation that predates the Internet by more than two millennia. A taxonomy can serve as more than a means to representing knowledge; its organization of information can also enable us to make information accessible and findable. Perhaps the most familiar examples of using a taxonomy for information access are the web directory that Yahoo! built in the mid-1990s [48] and the Open Directory Project, shown in Figure 2.4.

As an interface metaphor, the directory offers users a key advantage over search—independently of whether search is implemented using set or ranked retrieval. Because a directory organizes content, it provides users with guidance toward potentially interesting subsets of a document collec-

FIGURE 2.4: The open directory project.

tion. Although search requires a user to communicate to a machine based entirely on a preconceived information need, directory-based navigation allows a user to elaborate that need progressively, learning from the available options.

However, the problems that taxonomies pose for knowledge representation come back to haunt us when we use them for information access. Information seekers need to discover that path to a piece of information in the same way that the taxonomist conceived it. This coincidence of understanding is the exception, rather than the rule, as Furnas and others [49] observed in a seminal paper on the "vocabulary problem." Indexers disagree among themselves, and each searcher's mental model of how information should be organized is unique. As we saw from cats being classified under technology in the Dewey Decimal Classification system, what is intuitive to an indexer may not be intuitive to a user or to other indexers.

There have been attempts to work around the rigidity of directories for information access. In particular, some directories augment the strictly hierarchical taxonomy with symbolic links that offer alternate pathways to those defined by the taxonomy [50]. These work-arounds, however, are exactly that: an attempt to patch a fundamental limitation of taxonomies. They do not substitute for the benefits of Ranganathan's hospitality of notation.

2.6   TAKE-AWAYS

- The set retrieval model offers information seekers expressivity and transparency, but even professionally trained users are unable to formulate Boolean queries effectively.
- Ranked retrieval addresses the usability challenges of set retrieval but at the price of losing the bright line of the set retrieval model, which is problematic for approaches like faceted search that depend on the notion of a result set.
- Web search engines have introduced a query-independent notion of document authority that is particularly well suited for hypertext collections because citation loosely implies endorsement.
- Taxonomies offer directory-based navigation as an alternative to search, but there is a vocabulary problem: consumers must seek information in the same way indexers organized it.

*   *   *   *

CHAPTER 3

Faceted Information Retrieval

"Man's mind, once stretched by a new idea, never regains its original dimensions."
—Oliver Wendell Holmes, Sr.

Now we can put together what we have learned in the previous two chapters. In the first chapter, we saw that faceted classification addresses some of the limitations of a taxonomy for knowledge representation. In the second chapter, we saw that both search and directory navigation have limitations as interfaces for information retrieval. In this chapter, we will see how we can build better interfaces for information retrieval by using faceted classification.

Before we discuss faceted search, we will consider its predecessors: parametric search and faceted navigation. Neither of these predecessors considers the textual aspect of documents—a concern we will defer until we discuss faceted search itself.

3.1   PARAMETRIC SEARCH

As our running example in this chapter, we will use a domain popular with faceted search researchers and practitioners: wine. The facets typically used to characterize wine include varietal (the type of grape, e.g., Merlot), vintage (the year in which a wine was produced), region, rating, price, vintner, and so on. How can we apply a faceted representation of the data to help someone (whom we will call the user) find a wine that meets his or her particular tastes?

A parametric search interface is essentially a Boolean search interface for a faceted content collection: it allows users to formulate queries by visually specifying a set of constraints on the facet values [51]. A query is typically an AND of ORs: values selected within a single facet are combined using a logical OR, whereas constraints associated with different facets are combined using a logical AND. The system responds to a query with the set of objects in the collection that satisfy it.

Let us use a concrete example to see how parametric search works. Consider a user interested in red wines from France with a rating of at least 90 and a price at most $10. Parametric search allows him or her to specify this query as the following set of constraints: {Varietal: Red, Region: France, Rating: ≥90, Price: ≤$10}. A parametric search interface, such as the one depicted in Figure 3.1, presents each facet for independent specification. Note that the simple interface pictured does not

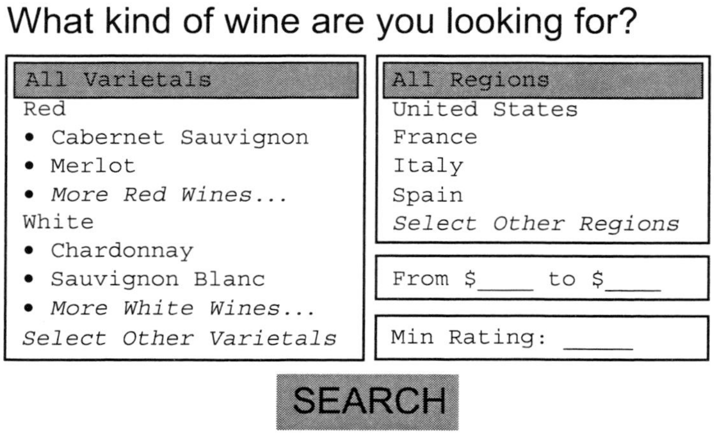

What kind of wine are you looking for?

All Varietals	All Regions
Red	United States
• Cabernet Sauvignon	France
• Merlot	Italy
• *More Red Wines...*	Spain
White	*Select Other Regions*
• Chardonnay	
• Sauvignon Blanc	From $\_\_\_\_$ to $\_\_\_\_$
• *More White Wines...*	
Select Other Varietals	Min Rating: _____

SEARCH

FIGURE 3.1: A simple parametric search interface.

enable OR-ing within a facet; we will discuss the interface challenge of allowing multiple selections from a facet in Section 7.4.

As should be clear from even this simple example, different kinds of facets lend themselves to different query metaphors.

For a facet that has nominal values (i.e., a list of enumerable categorical values), it makes sense for the user to see a list of options and select one or more of these individually. If the list is large (e.g., a wineries facet), then further effort is necessary to avoid information overload, as we will discuss in detail in Section 6.3.

For a hierarchical facet, such as the varietal facet in our example, the user might select a nonleaf value, such as Red, or a leaf value, such as Merlot. A user might see both leaf and nonleaf options simultaneously or might work top-down through the hierarchy.

For numerical facets, such as price or rating, the user most likely wants to select a range, possibly unbounded on one side. In the visualized interface, there is no guidance as to what are reasonable bounds; a more sophisticated interface might provide such guidance.

We will consider interface issues specific to faceted search in the Chapter 7. Nonetheless, we raise some of them now to emphasize how interface design is a critical concern in any application that queries against a faceted collection.

Importantly, parametric search is a form of set retrieval: it offers a subset of Boolean search functionality, albeit over facets rather than unstructured text. Like Boolean search over text, it suffers from the problem that users struggle to formulate their queries. They run into a "million or none" problem: underspecified queries return too many results, whereas overspecified queries return no results. Parametric search offers expressivity, but it does not offer users guidance through the space of possible queries.

3.2   FACETED NAVIGATION

Faceted navigation fills in the piece that is missing in parametric search: guidance. Parametric search requires that the user express an information need as a query in one shot, making selections across all facets of interest. In contrast, faceted navigation allows the user to elaborate a query progressively, seeing the effect of each choice in one facet on the available choices in other facets.

To illustrate the difference, let us return to our previous wine example. A user might want to start by establishing a budget of $10. This selection constrains the other facets—for example, there are no French wines for less than $10. Further selections in the varietal and region facets constrain the options in the unspecified facets, for example, selecting Spain as a region eliminates Sauvignon Blanc as an option for the varietal facet. Eventually, the user chooses to see all results that match the specified constraints. The faceted navigation interface in Figure 3.2 illustrates this sequence of steps.

Faceted navigation delivers an experience of progressive query refinement or elaboration. From a user's perspective, faceted navigation eliminates the "dead ends" that can result from selecting unsatisfiable combinations of constraints among the facets. In fact, most combinations of facet

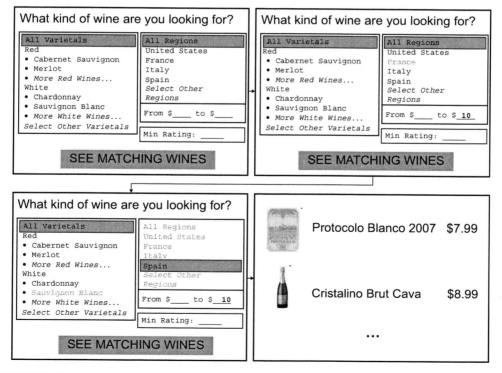

FIGURE 3.2: Faceted navigation interface.

values are unsatisfiable because the set of satisfiable combinations is typically a sparse subset of the set of all possible combinations. Thus, faceted navigation addresses the "million or none" problem of parametric search.

Not all kinds of facets, however, lend themselves to faceted navigation interfaces. For example, in the interface depicted above, the user could still arrive at a dead end by selecting wines for less than $1. A more sophisticated user interface might avoid this possibility, for example, by dividing numerical values into discrete ranges. In general, faceted navigation only makes sense for facets whose values can be presented through tractable choice lists. In the third part of this book, we will discuss techniques that help determine or present such tractable choice lists for facets that have large numbers of possible values.

But how do we handle textual data? That question leads us to the next section—and the subject of this book: faceted search.

3.3   FACETED SEARCH

We finally arrive at faceted search, the topic that motivated this lecture. After so much introduction, a small section on faceted search may seem anticlimactic! The goal of this section, however, is to provide a minimal definition of faceted search that will serve as the basis for the rest of this lecture.

In the previous chapter, we considered information retrieval approaches designed for searching text collections. In contrast, the parametric search and faceted navigation approaches described are oblivious to unstructured text and instead assume documents to be collections of values in a faceted classification system.

In practice, most of the document collections that interest us are semistructured: a typical document contains a combination of unstructured text and structured attributes. The structured attributes are sometimes called metadata. The Greek μετά (*meta*) means "with" or "adjacent" but typically means "about" in its common use as a prefix. Indeed, a document's metadata consists of structured attributes about the document that are, at least in a logical sense, stored with the document.

When this structured content conforms to a faceted classification system, we can combine text search, applied to the unstructured text content, with faceted navigation of the structured content. This approach is the essence of faceted search.

Let us go back to our wine example, but this time assuming a richer document model where each wine is also associated with descriptive text. The example in Figure 3.3, taken from an Endeca demonstration application, illustrates how faceted search combines a text-oriented search with faceted navigation [52].

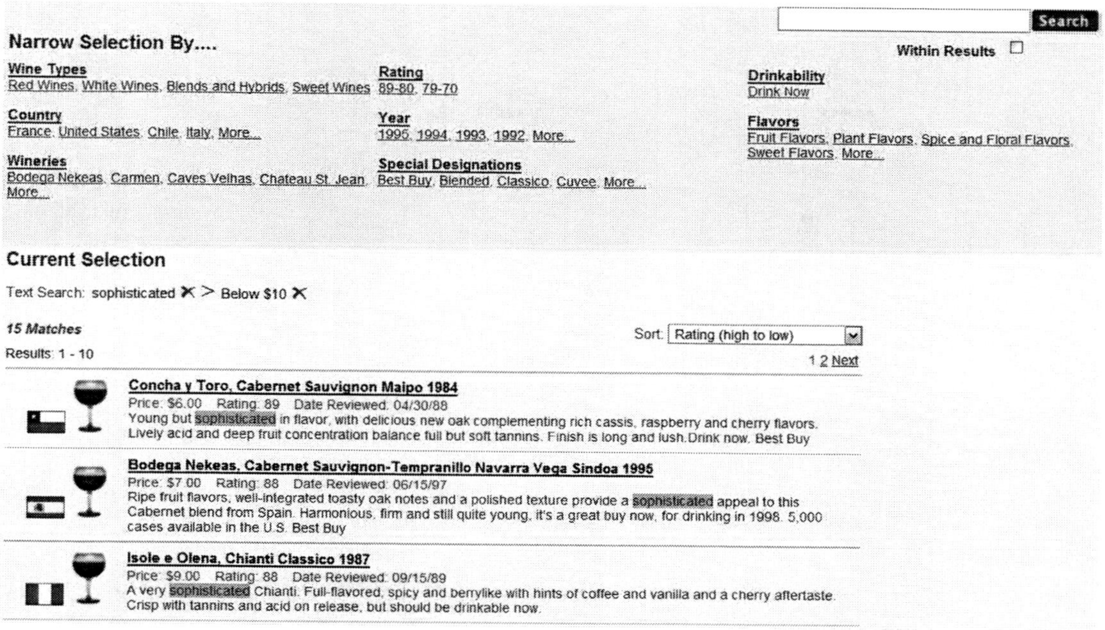

Narrow Selection By....

Wine Types
Red Wines, White Wines, Blends and Hybrids, Sweet Wines

Rating
89-80, 79-70

Drinkability
Drink Now

Country
France, United States, Chile, Italy, More...

Year
1995, 1994, 1993, 1992, More...

Flavors
Fruit Flavors, Plant Flavors, Spice and Floral Flavors, Sweet Flavors, More...

Wineries
Bodega Nekeas, Carmen, Caves Velhas, Chateau St. Jean, More...

Special Designations
Best Buy, Blended, Classico, Cuvee, More...

Search
Within Results ☐

Current Selection

Text Search: sophisticated ✗ > Below $10 ✗

15 Matches

Results: 1 - 10

Sort: Rating (high to low)

1 2 Next

Concha y Toro, Cabernet Sauvignon Maipo 1984
Price: $6.00   Rating: 89   Date Reviewed: 04/30/88
Young but sophisticated in flavor, with delicious new oak complementing rich cassis, raspberry and cherry flavors. Lively acid and deep fruit concentration balance full but soft tannins. Finish is long and lush. Drink now. Best Buy

Bodega Nekeas, Cabernet Sauvignon-Tempranillo Navarra Vega Sindoa 1995
Price: $7.00   Rating: 88   Date Reviewed: 06/15/97
Ripe fruit flavors, well-integrated toasty oak notes and a polished texture provide a sophisticated appeal to this Cabernet blend from Spain. Harmonious, firm and still quite young, it's a great buy now, for drinking in 1998. 5,000 cases available in the U.S. Best Buy

Isole e Olena, Chianti Classico 1987
Price: $9.00   Rating: 88   Date Reviewed: 09/15/89
A very sophisticated Chianti. Full-flavored, spicy and berrylike with hints of coffee and vanilla and a cherry aftertaste. Crisp with tannins and acid on release, but should be drinkable now.

FIGURE 3.3: Faceted search interface.

This time, the user does not start by using the facets but rather performs a free-text search for wines that contain the word "sophisticated" in their description. The user then uses the price facet to narrow these results to those wines less than $10. The system returns the 15 wines that match these filters, sorted by rating (another facet). The user can further refine this query by selecting values from other facets, such as type, country, and so on.

Like faceted navigation, faceted search eliminates choices among the facet values that would lead to dead ends. For example, the rating range of 90–100 is not available as a refinement option— evidently you have to cut some corners when you are looking for inexpensive, sophisticated wine!

So there you have it: faceted search. To quote a famous Peggy Lee song, is that all there is? You might wonder why you have picked up an entire book about faceted search only to find it summed up in a couple of pages when you are not quite halfway through.

As it turns out, faceted search is much like chess—it takes only minutes to grasp the rules but years to get the hang of playing the game well. The second part of this book reviews what academic researchers and industry practitioners have done with faceted search over the past decade. The third part addresses the implementation and design challenges of building faceted search systems. Whether your interests are academic or practical, we aim to offer a useful overview of what has been accomplished so far, and to identify the key directions that remain to be explored.

3.4 TAKE-AWAYS

- Parametric search offers a visual Boolean search interface for faceted content collections.
- Parametric search suffers from the problem that users may not able to formulate Boolean queries **effectively**.
- Faceted navigation provides the guidance missing in parametric search, allowing users to elaborate queries progressively.
- Faceted navigation allows users to see the impact of each incremental choice in one facet on choices in other facets.
- Faceted search combines faceted navigation with text search, allowing users to access semi-structured content collections.
- Faceted search provides support for discovery and exploratory search, areas where conventional search fall short.

* * * *

PART II
Research and Practice

The second part of this book reviews some of the most significant contributions to the development of faceted search, both from academic researchers and industry practitioners. While we cannot hope to be exhaustive, we aim to cover the most important milestones in the past decade, during which faceted search has emerged as a mainstream technique for information access.

CHAPTER 4

Academic Research

"The task of a university is the creation of the future."

—Alfred North Whitehead

In the previous chapters, we followed a logical progression from taxonomies to facets to parametric search that ultimately arrived at faceted search. Now we will begin to explore the subject of faceted search in depth.

In this chapter, we consider some of the most significant academic contributions to the subject. Although no enumeration of such work can hope to be comprehensive, we hope that the research examples we provide give you a taste of the evolution of research on faceted search. The following chapter will consider commercial applications of faceted search.

4.1 DYNAMIC QUERIES AND QUERY PREVIEWS

It is hard to ascertain what was the earliest formal work on faceted navigation, but the two strong candidates are the work on dynamic queries and on query previews at the University of Maryland and the work on view-based search at the University of Huddersfield. Neither research effort described its work in terms of "facets"—but, then again, neither did Ranganathan. Nonetheless, it is clear that these two efforts in the 1990s catalyzed today's interest in faceted search.

In "Dynamic Queries for Visual Information Seeking" [53], University of Maryland researcher Ben Shneiderman defines dynamic queries as "interactive user control of visual query parameters that generate a rapid . . . animated visual display of database search results." He emphasizes the speed and interactivity of such an interface, as contrasted with the alternatives of set retrieval using database querying languages like SQL [54]. In particular, dynamic queries enable users "to discover quickly which sections of the multidimensional search space are densely and sparsely populated, where there are clusters, exceptions, gaps, or outliers, and what trends there are in ordinal data."

Figure 4.1 shows the FilmFinder prototype built by Shneiderman and Christopher Ahlberg to enable exploration of a movie database [55]. The graphic design combines various interface elements, including parametric search over a faceted information space. The rapid feedback in

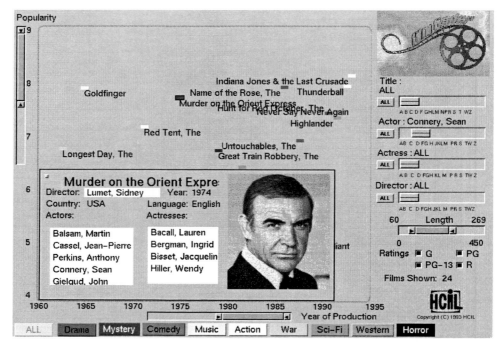

FIGURE 4.1: Dynamic queries using FilmFinder. (Human-Computer Interaction Lab, University of Maryland. Used with permission.)

response to incremental query modifications helps overcome the aforementioned difficulty of formulating queries in a parametric search interface.

Visual interfaces are certainly more friendly to nonexpert users than command-line interfaces that require formal query expressions. Moreover, rapid feedback of dynamic queries overcomes much of the frustration that can arise from parametric search over a sparsely populated space of possible combinations of values. Nonetheless, the feedback they offer users is reactive, not proactive. Users are still able to select unsatisfiable combinations.

Shneiderman, together with colleagues Khoa Doan and Catherine Plaisant, addressed this problem in later work on query previews [56]. Query previews "prevent wasted steps by eliminating zero-hit queries." In other words, query previews replace parametric search with faceted navigation.

Figure 4.2 shows how query previews are used in an implementation of faceted navigation for a collection of environmental data from the National Aeronautics and Space Administration (NASA). The left and right show a before and after: the selection of North America from the geography facet eliminates values like Sea Ice from the attribute facet and 1983 from the year

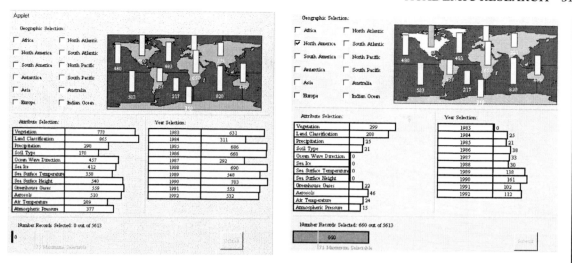

FIGURE 4.2: Query previews on NASA data. (Human-Computer Interaction Lab, University of Maryland. Used with permission.)

facet. Moreover, the dynamically updated counts allow users to see an overview of the distribution of values over the selected subset of documents.

4.2   VIEW-BASED SEARCH

Across the pond, Steven Pollitt at the University of Huddersfield independently led a project on faceted navigation, although he also did not call it by that name. Rather he and his colleagues described their approach as "view-based search."

In "View-based search systems—progress toward effective disintermediation" [57], Pollitt et al. describe their goal as "enabling end-users of retrieval systems to make effective use of databases without the assistance of search intermediaries."

They propose "view-based searching" as an approach in which:

> The user is now provided with much more opportunity to examine the database and to apply powerful searching as we can simultaneously present several views and employ them to examine the contents of the database by refinement and expansion of the different searching elements. . . . The paradigm shift to searching through reciprocally refining views has moved the interaction between user and machine further towards the subject matter and away from the operation of the system.

What Pollitt calls "views" are what we call facets. Indeed, view-based search is not "search" at all but rather is an implementation of faceted navigation. Pollitt's work may be the earliest example of hierarchical faceted navigation—that is, faceted navigation where at least some of the facets correspond to taxonomies that include nonleaf values.

Figure 4.3 depicts the HIBROWSE implementation of view-based search by Pollitt et al. applied to a collection of documents from Lexis-Nexis.

Both view-based search and the previously described work on query previews suffer from two limitations.

The first limitation is that they support faceted navigation but not faceted search. The lack of free-text search turns out to be a deal-breaker in most information access interfaces—especially for today's users, a population that has grown up with ubiquitous search.

The second limitation is that they only show the user either a default set of facets or facets that the user has explicitly selected. The system does not automatically add views or delete views to reflect the query context. As we will discuss at greater length in the third part of the lecture,

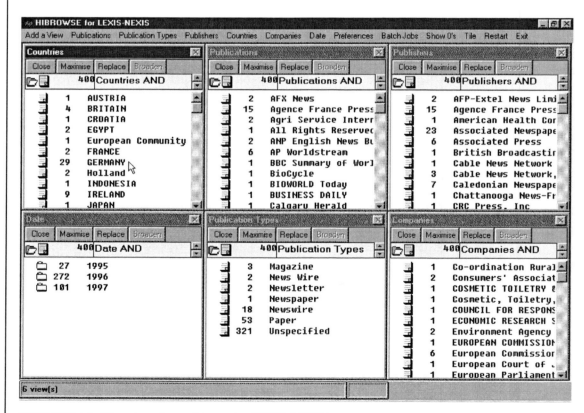

FIGURE 4.3: View-based searching using HIBROWSE.

this limitation places an unreasonable burden on users when there are even a moderate number of facets.

4.3   THE FLAMENCO PROJECT

Perhaps the person most associated with faceted search is Marti Hearst. In the mid 1990s, Hearst, then a researcher at the Xerox Palo Alto Research Center, joined a group developing Scatter/Gather, a cluster-based approach to browsing large document collections [58]. Scatter/Gather clusters documents into semantically coherent groups and presents summaries of the groups to the user. Most clustering systems today hark back to this seminal work, shown in Figure 4.4.

Clustering shares some common interface goals with faceted search—in particular, it enables the user to explore a collection through interaction and a form of query preview. As we can see, however, the Scatter/Gather work assumes that documents only contain unstructured text. This minimal data model limits the power of an exploratory interface.

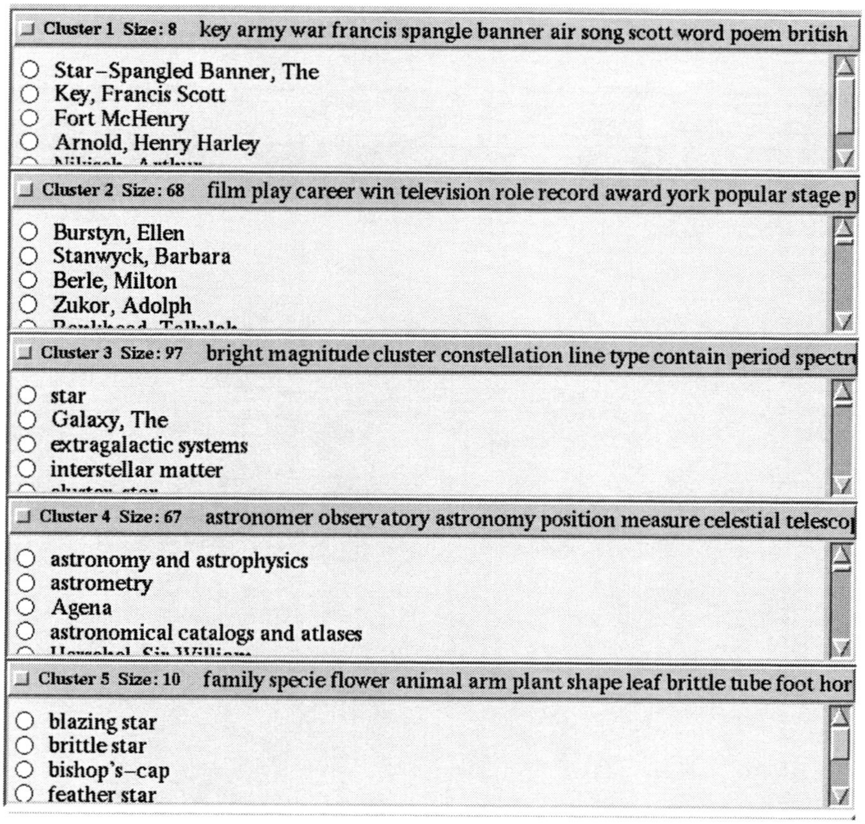

FIGURE 4.4: Scatter/Gather: results for "star."

Subsequent to the Scatter/Gather work, Hearst joined the faculty at the University of California, Berkeley, and embarked on a project that focused entirely on faceted search: FLexible information Access using MEtadata in Novel COmbinations, better known as Flamenco. Hearst describes its agenda as follows [59]:

> Our end goal is to develop a general methodology for specifying task-oriented search interfaces across a wide variety of domains and tasks. We suggest that rich, faceted metadata be used in a flexible manner to give users information about where to go next, and to have these suggestions and hints reflect the users' individual tasks.

The Flamenco project represents almost a decade of work on developing faceted search tools and performing usability studies with them. Its centerpiece is an open-source faceted search system that supports hierarchical facets [60].

In addition to building tools to enable faceted search, Hearst and her colleagues have researched issues of interface design [61] and automating metadata creation [62, 63]. Hearst has also performed usability studies comparing faceted search to the clustering approach of her earlier Scatter/Gather work [64].

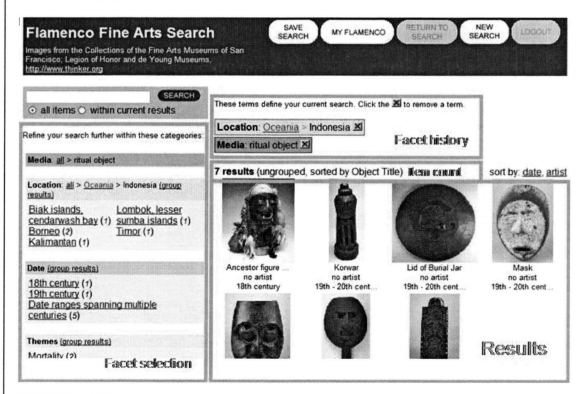

FIGURE 4.5: Flamenco project.

Figure 4.5 illustrates an example Flamenco application: a faceted search system over a collection of images from fine arts museums in San Francisco.

4.4 RELATION BROWSER

Whereas Hearst and her colleagues have pursued the Flamenco project at UC Berkeley, Gary Marchionini has led a parallel effort at the University of North Carolina, Chapel Hill, called the Relation Browser [65]. This project, originally developed for the US Bureau of Labor Statistics, aims to improve on the available mechanisms for searching and navigating the bureau's web site by offering a preview-oriented interface. Unlike the Flamenco work, the Relation Browser allows users to quickly explore a document space using dynamic queries issued by mousing over facet elements in the interface.

Because of their consistent focus on federal statistics as a domain, Marchionini and colleagues have been able to iteratively improve their interface based on several years of user studies [66, 67].

Figure 4.6 shows the interface for RB++, the most recent version of the Relation Browser at the time of this writing. Its features include the use of graphic bars to indicate the counts associated with facet values. The Relation Browser allows users to quickly explore a document space using dynamic queries issued by mousing over facet elements in the interface.

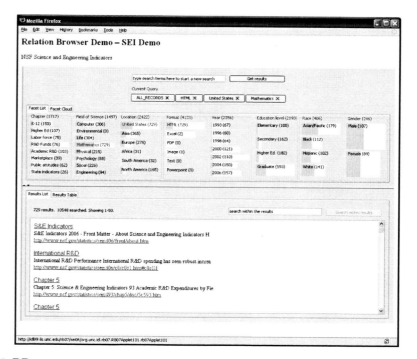

FIGURE 4.6: RB++.

4.5 mSPACE

University of Southampton researcher mc schraefel (yes, her name is case-sensitive) and colleagues describe the mSpace project as "an interaction design to support user-determined adaptable content and describe three techniques which support the interaction: preview cues, dimensional sorting and spatial context" [68]. Not surprisingly, these goals have much in common with the work described in the previous sections. The mSpace work, however, has placed more emphasis on user interface details.

For example, the mSpace Classical Music Explorer [69], shown in Figure 4.7, uses a spatial multicolumn layout and allows users to both select and rearrange the ordering of facets. The interface also offers audio previews of exemplar documents (in this case, classical music works), and the researchers studied the effect of varying the number of available exemplars on user experience.

4.6 PARALLAX

The last academic research project is Parallax [70]. David Huynh initially developed Parallax at MIT, as part of David Karger's SIMILE project, but has continued this work at Metaweb as an interface for Freebase, a collaborative knowledge base of structured data [71]. The Parallax interface

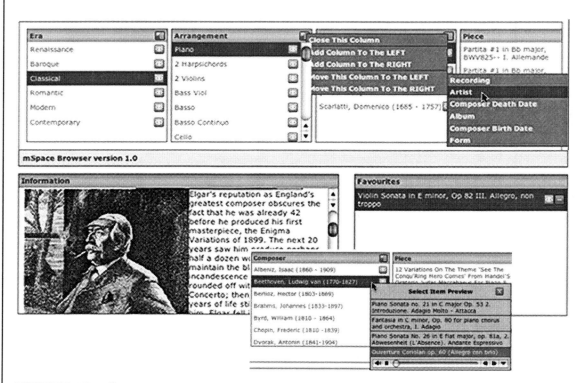

FIGURE 4.7: mSpace.

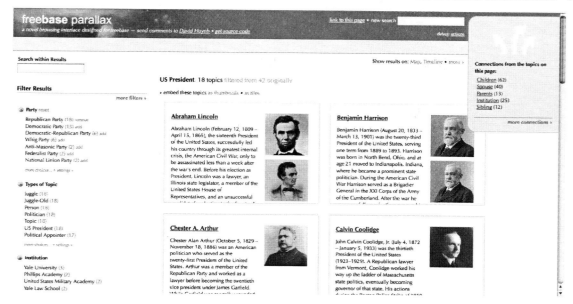

FIGURE 4.8: Freebase Parallax.

offers "set-based browsing" that extends faceted search to shift views between related sets of entities, a semantic web approach similar to Endeca's "record relationship navigation" [72].

Figure 4.8 illustrates the Parallax interface as applied to Freebase, specifically to an exploration of US presidents. As the user views a set of results, Parallax not only provides filters based on facets associated with those results but also uses other relations in the ontology to provide connections (see the upper right) to related sets that have their own facets. Thus, a user can filter to presidents of a particular party, shift views to their parents, and then view their professions, nationalities, and so forth. Like mSpace, Parallax pays particular attention to visual design and user interface details.

Strictly speaking, Parallax is not a faceted search system but a semantic web (or set-based) browser. Because it supports a more general ontology, Parallax allows for the richer set of relationships discussed in Section 1.3. This power, however, comes at a price: as we discuss in Section 6.6, implementing this generalization of faceted search raises challenges, both for computation and usability. Nonetheless, Parallax is a significant step toward making the semantic web explorable, using many of the same techniques that have made faceted search successful.

4.7   TAKE-AWAYS

- Faceted search has been an active area of academic research for about a decade.
- The early work on dynamic queries, query previews, and view-based search showed the potential for faceted navigation.

- The Flamenco project at UC Berkeley was and continues to be the most visible research project focusing on faceted search.
- The Relation Browser project at UNC offers a deep study of faceted search in the domain of statistical data.
- The mSpace project emphasizes the user interface aspects of faceted search, a particular concern for unfamiliar users.
- Parallax generalizes faceted search to the general ontologies associated with the semantic web.

CHAPTER 5

Commercial Applications

"Ideas are refined and multiplied in the commerce of minds."

—Gaston Bachelard

As the previous chapter shows, academic research has led the way in advancing our understanding of faceted search as a technique to move beyond the limitations of previous approaches to information retrieval. Nonetheless, the biggest success stories of faceted search come from its use in commercial applications. Over the past decade, faceted search has gone from being an esoteric research topic to a ubiquitous feature of commercial search applications. This brief chapter makes no attempt to enumerate the many companies that have built such applications but rather highlights some of the most visible milestones in the commercialization of faceted search.

5.1 ENDECA

Endeca was founded in 1999 with a vision of delivering faceted search, branded as Guided Navigation, to enterprises [73]. Endeca is most known for providing faceted search for e-commerce sites (e.g., Wal-Mart and Home Depot) but also applies the same technology to other domains, including manufacturing, publishing, financial services, and government. Endeca is generally credited with evangelizing faceted search as a core feature for ecommerce and site search applications.

Figure 5.1 shows an example of an Endeca-powered application: the search interface for the Triangle Research Libraries Network [74], which built on an earlier effort at North Carolina State University [75]. The faceted search interface allows users to explore a semistructured catalog that associates each book with faceted metadata and descriptive text. TRLN [76] documented this effort and has also performed user studies.

5.2 eBAY

eBay, founded in 1996, is "the world's largest online marketplace" [77]. Although eBay has expanded beyond its original focus, it is still primarily known as an online auction site.

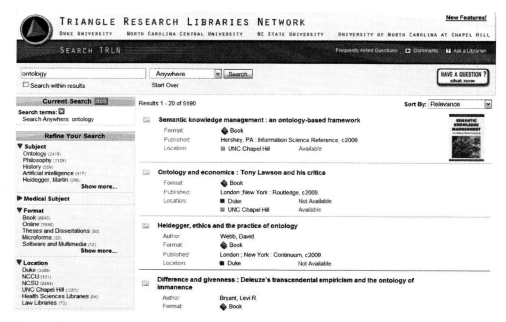

FIGURE 5.1: Triangle research libraries network.

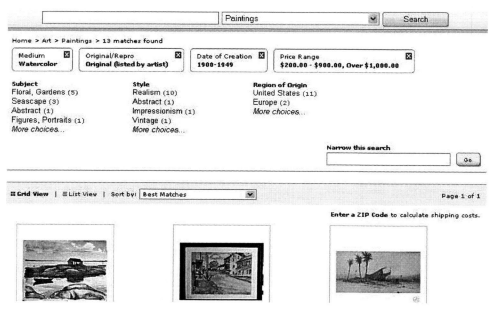

FIGURE 5.2: eBay Express.

In 2006, eBay embarked on a project called eBay Express, a web site designed to work more like a typical shopping site rather than one focused on auctions [78]. They shut the project down two years later but folded some of its features into their main site [79].

Figure 5.2, courtesy of Marti Hearst (whom eBay enlisted as a consultant for this effort [80]), illustrates the faceted search interface used by eBay Express.

5.3   AMAZON

Founded in 1994, Amazon.com is the largest online retailer in the United States [81]. Originally focused on books, Amazon.com has diversified to include a wider variety of product types, as well as services unrelated to its product sales.

In 2002, Amazon unveiled "Project Ruby," an experimental faceted search site for its multistore apparel department [82]. Project Ruby allowed shoppers to browse by store, product category, and brand. Over subsequent years, Amazon extended faceted search to all of its product categories. Figure 5.3 illustrates Amazon's faceted search interface for apparel at the time of this writing.

FIGURE 5.3: Results for "shoes" at Amazon.com.

5.4 OPEN SOURCE

Although commercial enterprise search vendors and major online retailers were the first to push faceted search into the mainstream, the open-source community followed shortly thereafter.

In 2006, CNET networks contributed Solr, a project it had originally developed in-house in 2004, to the Lucene project, one of the most widely used open-source libraries for search [83]. Although Solr offers several enhancements to the Lucene platform, the main one is the introduction of faceted search [84].

Another notable open-source use of faceted search is as a component of Drupal, an open-source content management platform [85]. Originally written by Dries Buytaert to implement a message board, Drupal became an open-source project in 2001. Drupal has since attracted a substantial user base and powers a number of high-profile web sites, including those of The New York Observer and Amnesty International.

Figure 5.4 demonstrates an implementation of Solr's faceted search on the CNET shopping site. As we can see, open-source implementations of faceted search look much like commercial ones—further evidence that faceted search has become mainstream.

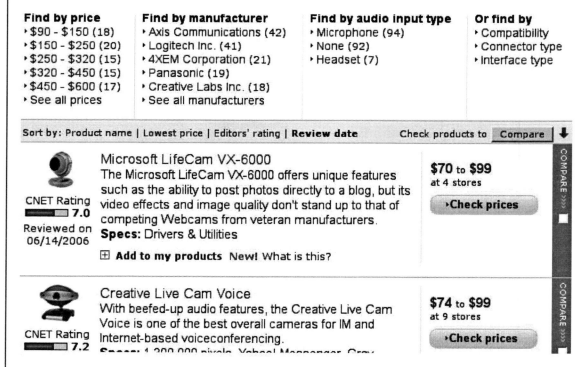

FIGURE 5.4: Results for "Webcam" at CNET shopping.

5.5   TAKE-AWAYS

- Faceted search has rapidly moved from an esoteric academic topic to a mainstream commercial technology.
- Vendors like Endeca have made faceted search a core feature of e-commerce and site search.
- Outside of web search engines, most major web companies—including Amazon and eBay—implement faceted search.
- The emergence of open-source options such as Solr and Drupal has made basic faceted search a commodity feature.

*   *   *   *

PART III
Practical Concerns

The third part of this book aims to be most useful to practitioners, but we also encourage researchers to look at these concerns as open problems ripe for further exploration. We split them into two broad categories: back end and front end.

The split is somewhat arbitrary, as there are many cross-cutting concerns. Nonetheless, we try to distinguish between the back-end concerns about determining what information to present and the front-end concerns of deciding how to present it. Sometimes the same strategy helps address both.

We hope that the breadth of considerations imparts a healthy respect for the challenges of building an effective faceted search system but not so healthy that it leaves the reader intimidated. As we have seen from the previous section, implementing basic faceted search, like playing rudimentary chess, is straightforward. Fortunately, it is possible to build valuable applications without being a grandmaster.

CHAPTER 6

Back-End Concerns

"The greatest challenge to any thinker is stating the problem in a way that will allow a solution."

—Bertrand Russell

In this chapter, we will consider some of the thorniest issues confronting the back-end developers of faceted search applications. These include systems issues of scale and efficiency, as well as information science issues, such as obtaining facets and overcoming vocabulary challenges. We also touch on the problem of handling multiple entity types.

We save the front-end concerns of design and usability for the next chapter.

6.1 SCALE

The value of a document collection is often correlated to its size; when it comes to information, more—or at least access to more—is better. Moreover, because faceted search enables exploration, a faceted search system potentially offers more utility for large document collections than a conventional search engine because it can make more of the information accessible in practice.

Unfortunately, the benefits of scale also incur costs, either in hardware requirements or in operational complexity. There are also efficiency concerns, but we will discuss those in the following section.

For faceted search systems, scale is itself a multifaceted concern. When we talk about scale, we primarily mean the following factors:

- Number of documents
- Number of facet values per document
- Searchable text per document

The precise storage requirements depend heavily on the particular data structures used to implement faceted search, but roughly speaking we assume two tables: an inverted index that maps facet values or searchable text to documents and a document table that maps documents to facet values. In such a representation, the size of a faceted search system is, roughly speaking, the product of the number of documents and the average "width" of a document—which corresponds to the number of facet values plus the amount of searchable text.

For small document collections, all of this information can be kept in main memory—an approach that, as we will see in the next section, has attractive efficiency implications. For larger collections, however, an in-memory approach may not be a realistic option.

As of the time of this writing, random-access memory (RAM) capacity for all but the most expensive servers is measured in gigabytes ($1G = 10^9$ bytes), whereas external storage capacity is measured in terabytes ($1T = 10^{12}$ bytes). Moreover, the cost of RAM accelerates sharply at the high end. Hence, at least for the foreseeable future, the cost of external storage is far lower than the cost of RAM, which means that most storage-intensive applications are likely to require some—if not most—of the storage to be external.

It is also possible to distribute storage across multiple servers. Such a distribution may be a scaling strategy: using distributed in-memory storage to avoid the accelerating cost of server RAM. Distributed storage may also be an external requirement imposed on an application developer for other reasons. Although the storage for a faceted search should scale linearly, some care is required to avoid inefficient computation—particularly network latency.

6.2   EFFICIENCY

Compared to the information retrieval approaches described in Chapter 2, faceted search is computationally demanding.

When a faceted search system processes a query, its first task is to determine the set of documents that satisfy the query constraints. This set retrieval is straightforward and can be accomplished efficiently using standard inverted index techniques, such as those described in Trevor Strohman's dissertation on the subject [86].

The subsequent task, however, is to show the facet values available for refining the set of results. In addition, faceted search applications often show the counts associated with these refinements. The task of computing refinements is significantly more demanding than that of computing the set of results.

There are two straightforward ways to compute refinements: top-down and bottom-up. A top-down approach leverages the inverted index, looking up each candidate facet value to compute the intersection of the documents assigned that value with the documents in the result set. A bottom-up approach iterates through the documents in the result set and then iterates through the facet values assigned to each document to accumulate them.

Both of these approaches are computationally expensive. The top-down approach requires computing a number of set intersections equal to the global number of facet values, whereas the bottom-up approach requires examining all facet value assignments for the documents in the result set.

The cost of either approach depends on the specific data structures used to represent the inverted index and documents and particularly on what information is stored in memory compared to that on disk. Bottom-up approaches can be particularly expensive if they access external storage because the storage is likely to be scattered across the disk; in contrast, top-down approaches are more likely to take advantage of locality.

Yonik Seeley, the creator of the open-source Solr described in the previous chapter, describes an approach that combines a top-down and a bottom-up approach to achieve better efficiency than either technique alone [87]. Unfortunately, commercial providers of faceted search systems do not disclose the proprietary techniques they use to achieve computational efficiency.

It is also important to decide how to measure efficiency, particularly when serving concurrent users. If the main concern is throughput, that is, the average number of queries processed per second, then most approaches make it possible to improve efficiency linearly by using additional servers that maintain copies of the document collection (although distributing the collection can complicate replication). In contrast, there are no straightforward approaches to reduce latency, that is, the time experienced by a user for a single query. As systems researcher David Clark tells us, "Bandwidth problems can be cured with money. Latency problems are harder because the speed of light is fixed—you can't bribe God" [88].

As noted in the previous section, it is possible to distribute storage across multiple servers. Such an approach can reduce query latency through intraquery parallelization, as can distributing query processing among multiple processes or threads on a single server. All of these distribution approaches, however, introduce complexity.

Distributed storage across multiple servers introduces operational complexity in the indexing process (particularly to maintain consistency), network latency that may outweigh the gains from distribution, and more complex query processing—including a possible need to issue multiple queries to obtain a response independent of the storage distribution [89].

Distributed query processing on a single server is more attractive operationally, but introduces its own complexity. Multiple processors or threads may contend for scarce computational resources, resulting in thrashing or blocking. Although the increasing availability of multicore servers makes this approach attractive, the current generation of software products is not yet able to fully exploit their computational power.

The other significant factor likely to affect efficiency is the frequency of data updates. Many faceted search systems assume that indexing documents is a relatively slow, offline process compared to the time-sensitive task of servicing queries while a user waits. In the presence of data updates, however, we have to consider how users experience update latency—that is, a lag between when an update arrives and when users experience its effect in query results—and the degradation of query processing efficiency that results from computational resources spent on the indexing.

A full discussion of how to optimize the computational efficiency of a faceted search system is beyond the scope of this book. What we hope this section makes clear is that the computation cost of faceted search is significantly more expensive and complex than that of ordinary search engines and that the number of variables that affect this cost is high enough that application developers should proceed with caution, testing early and often.

6.3   INFORMATION OVERLOAD

The computational challenges discussed in the previous two sections are serious, but a challenge that is (at least) as serious is the scarcity of the most valuable resource of all—the user's attention. As Herbert Simon, the Nobel Laureate who pioneered the study of bounded rationality, says, "a wealth of information creates a poverty of attention and a need to allocate that attention efficiently" [90]. The wealth of information offered by faceted search systems threatens to overwhelm users and thus requires us to prioritize what information we show to users efficiently, thus optimizing the allocation of users' attention.

To some extent, the challenge of information overload falls under the jurisdiction of front-end user interface design, which we will discuss in the following chapter. Some concerns, however, have as much to do with what information is presented as with how it is presented to users. We discuss the former concerns in this section.

There are essentially two factors that can lead to information overload in a faceted search system: the number of facets and the number of facet values. Given the limitations of screen real estate and human attention, we cannot always show users all facets and all facet values. There are ways to prune both.

The number of facets reflects the number of ways a document can potentially be classified. In theory, there is no limit to the number of facets; there are infinitely many potential taxonomies to classify a document collection. In practice, of course, the number of facets is finite, but it may be quite large.

There is also the issue of dependence among facets. In his colon classification, Ranganathan intended the facets to be independent of one another; in most practical settings, however, we cannot assume such independence. City, state, and country may exist as three distinct facets rather than a single hierarchical facet; language and nationality are highly correlated and yet clearly distinct facets; and yet other facets, such as medical subject within a library catalog, may only apply to a subset of a document collection. At best, designing a faceted classification scheme with independent facets requires extraordinary effort on the part of information architects; at worst, it is an impossible task because such a set of independent facets would not match cleanly to the way users conceive of the information space. Either way, we cannot require that facets be independent of one another.

Rather, we need to face the possibility of a large number of heterogeneous, interdependent facets. We cannot present all of these to users all the time—not so much because of the computational costs (although these could be significant) but because the information would overwhelm users. Instead we must attempt to determine the most valuable facets to present to users.

Here are a few techniques from the conventional wisdom and the research literature [91, 92]:

- Favor facets with high coverage in the result set—in particular, favor facets with values assigned to all documents in the results set over those only assigned to a small subset.
- Favor facets that yield a high-entropy distribution of values in the result set. For example, a facet with ten values (e.g., document language) is more valuable if the distribution is uniform (i.e., 10% of the documents associated with each value) than if the distribution is highly concentrated (e.g., 99% of the documents are in English). In fact, we can think of favoring high coverage as a special case of favoring high entropy, since uncovered documents will all share the same "unassigned" value for the facet, resulting in a low entropy for the distribution.
- Consolidate facets that contain common values. For example, document facets might include authors, editors, contributors, and participants, and it is likely that their sets of values will overlap. It may thus make sense to consolidate these in a single facet—particularly if the distinction between these facets is arbitrary (e.g., an document's editor may sometimes be listed as an author).

The second information overload challenge comes from facets that have large numbers of values. Just as there is a limit to the number of facets that a system can present to a user, there is a limit to the number of values from any given facet.

Ideally, a facet with a large number of values is hierarchical, as in Ranganathan's colon classification. For a hierarchical facet, the system can disclose refinement options progressively, only showing the children of the root or of a selected facet value. As long as no value in the hierarchy has a large number of immediate children, this approach avoids the problem of presenting a large number of values simultaneously. Enforcing such a constraint, however, may result in a deep hierarchy—which leads to its own usability issues. We must bear in mind that facets are supposed to help users, not to distract them with a impenetrable hierarchy!

Moreover, it is not always possible or practical to arrange facet values into a hierarchy. Arranging a collection of values into a meaningful hierarchy may be a significantly more laborious task than obtaining those values, particularly if the values have been obtained through automatic means or social tagging [93]. Moreover, some facets, such as names of people, may not be amenable to hierarchical organization. An arbitrary hierarchy may be worse than having no hierarchy at all: imagine organizing people into a hierarchy base on the month, date, and hour of their birth! Even without considering such absurd possibilities, we have to be careful not to be so bent on hierarchical organization that we forget our motivation for using facets in the first place—to overcome the limits of strict hierarchical organization.

In practice, we have to accept that some facets will have large numbers of values that cannot be arranged in a neat hierarchy.

Here are some techniques for mitigating overload from a nonhierarchical facet with large number of values:

- Show only values with the highest frequency in the result set.
- Show only values that occur more frequently in the result set than in the overall collection. This variation of the previous strategy is similar to using tf*idf for information retrieval.
- Create a hierarchy of values based on string prefixes (e.g., A–D, E–H), numerical ranges, or some other means of dividing the range of values.
- Cluster values based on a similarity measure, that is, infer a hierarchy based on correlation statistics among the values, either in the result set or in the overall collection.

Large numbers of facets and facets with large number of values pose significant information overload challenges, and the best solution, where possible, is to limit the number of facets and the number of values per facet. When such limitations are not possible or practical, the above techniques can help handle and mitigate information overload.

6.4   AVAILABILITY OF METADATA

Faceted search presumes that a collection of documents has been organized based on a faceted classification system. Unfortunately, it is often the case that people have collections of documents that have not been classified at all—collections of what is colloquially known as unstructured text. Can faceted search help in such cases?

The short answer is no: faceted search requires documents with faceted metadata. Without the metadata, all we have is text search.

The longer answer is that there are various techniques for enriching unstructured text to obtain faceted metadata. Many of these techniques fall under the rubric of text mining. Text mining is a rich field, and the interested reader is directed to more comprehensive literature on the subject [94, 95].

The following are a few general strategies:

- Exploit latent metadata, such as document source, type, length, and so on. Often this information is readily available.
- Use rule-based or statistical categorization to classify documents into predetermined categories. Multiple independent categorizations can yield different facets.
- Use an unsupervised approach like terminology extraction to obtain a vocabulary of terms from the document collection.

In short, faceted search requires facets to be useful. Obtaining such facets from unstructured text is an option that may be amenable to automation but will likely require some amount of manual intervention. In particular, human postprocessing may be needed to cleanse automatically extracted facet values, as well as to arrange them into hierarchies or split them into multiple facets.

An approach we have not discussed is crowd-sourcing—that is, getting other people to provide metadata for you. There are a number of ways to leverage "human computation," ranging from paid services like Amazon's Mechanical Turk [96] to mining search logs for potential facet values [97]. There are even "games with a purpose" [98] designed to help label images; perhaps they will also prove useful for obtaining faceted metadata.

6.5   THE VOCABULARY PROBLEM

In 1987, Furnas et al. published a seminal paper entitled "The Vocabulary Problem in Human-System Communication," (also discussed in Chapter 2) in which they demonstrated how "the keywords that are assigned by indexers are often at odds with those tried by searchers." Their proposed solution is an "adaptive indexing" approach that collects word usage data from user behavior [99].

The vocabulary problem is both a motivation and a challenge for faceted search. Faceted refinement options offer the user guidance, thus mitigating the user's unfamiliarity with the indexer's vocabulary. At the same time, faceted search systems rely not only on matching user-specified search terms against document text but also on users understanding refinement options presented to them.

The first part of this challenge reflects the inherent lack of precision of user-entered search queries and is a larger topic in information retrieval. The interested reader is encouraged to consult a textbook on information retrieval to learn more about the variety of query expansion and relevance feedback techniques aimed at addressing this problem [100].

The second part of this challenge is more specific to faceted search—or at least to refinement-oriented information retrieval interfaces. For users to make effective use of refinement options, the refinements must offer users what Pirolli et al. [101] call "information scent": cues that indicate to users the value, cost of access, and location of available information content. It is thus important that a user, considering the available set of refinement options, see at least one option that offers the needed information scent and ideally does not see two or more that create confusion as to what is the best path to follow.

Faceted search systems can increase the information scent associated with a refinement option by offering previews of the content associated with that selection—and several of the research projects discussed in Chapter 4 show examples of how to do so. Ideally, the previews should be concise and yet should clearly indicate how the selection changes the current query state.

To avoid confusing users with too many plausible refinement options, system designers have at least two options: cleaning up the facet values to avoid duplication and clustering similar facet values when they are presented to users. These two options are not exclusive, and ideally a system should do both.

6.6   MULTIPLE ENTITY TYPES

The last issue we consider in this chapter is that of document collections with multiple entity types and relationships between entities of different types.

For those unfamiliar with this concept, let us consider a concrete example: a digital library of articles, where each article in the library has one or more authors. There are facets that describe articles, such as subject, year of publication, language, and so on. We can think of the authors themselves as constituting a facet that describes articles.

Where we run into problems is when we want to characterize the authors using facets, such as nationality, affiliation, and so on. Specifically, what happens when a user wants to explore the collection of articles based not only on the facets that apply to articles but also based on the facets that apply to authors? And what if an article can have multiple authors, as is often the case?

A conventional approach to faceted search assumes a denormalized model (we use the term a bit loosely; perhaps "un-normalized" is more appropriate): there are only two kinds of objects, documents and facets. If authors constitute a facet, then they cannot be documents—and hence they cannot themselves be assigned facets.

This denormalized model has significant implications for what kinds of queries users can make. For example, users cannot request articles by authors from the United States unless this nationality is represented as a facet assigned to articles—rather than to authors.

Moreover, because an article can have multiple authors, denormalization is a lossy representation. For example, the set of articles authored by someone from the United States who is affiliated with Carnegie Mellon University is not the same as the intersection of articles authored by someone from the United States and the set of articles authored by someone affiliated with the Carnegie

Mellon University—because an article may have two authors, each of whom satisfies one of the criteria.

To preserve these distinctions, we need to maintain both articles and authors as documents. Thus, articles can have facets and authors can have their own facets. An author facet value assigned to an article acts a reference to an author document.

Such a semantic web approach, like those proposed by Endeca and by David Huynh and described in Section 4.4, allows users to perform faceted search on multiple entity types simultaneously and to perform join operations between entity types as part of a query.

We note, however, that this approach is complex, both in terms of managing computational requirements and in designing its user experience. Generalizing faceted search to the general ontologies of the semantic web is still an active research area.

6.7   TAKE-AWAYS

- The size of a faceted search system depends on the number of documents, the number of facet values per document, and the amount of searchable text.
- Small document collections may be kept in main memory; for larger collections, external or distributed storage is necessary.
- Faceted search requires more computation than conventional search to compute facet refinements and counts.
- It is usually possible to increase throughput by adding hardware, but harder to reduce latency.
- The frequency of data updates can have significant impact on query latency and update latency.
- Large numbers of facets or facet values require particular attention to avoid information overload.
- Faceted search requires facets to be useful, which may require using text mining approaches to obtain faceted metadata.
- Faceted search, although designed to address the vocabulary problem, requires care to avoid exacerbating it.
- Faceted search assumes a denormalized data model and creates special challenges when applied to a collection with relationships among multiple entity types.

CHAPTER 7

Front-End Concerns

"A common mistake that people make when trying to design something completely foolproof is to underestimate the ingenuity of complete fools."

—Douglas Adams

In the previous chapter, we focused on the back-end challenges of determining what information to present to users and how to compute that information. In this chapter, we look at the complementary front-end challenges of presenting that information in a way that optimizes user experience. Although we cannot hope to be exhaustive, our aim is to address some of the most common and thorniest user interface challenges that arise in the design of faceted search systems.

7.1 WHERE AND WHEN TO PRESENT FACETS

A faceted search system responds to a query by returning a set of documents intended to match the search terms, and also provides a set of facets that offer the user directions for query refinement. This faceted search approach, however, does not specify how those two sets (the results and the facets) should be presented to the user. An application may choose to present both sets in the same view or may initially present only one set, that is, the results *or* the facets. Is there an optimal way to make these choices?

The short answer is no: different applications and user needs motivate different design choices. What we can do, however, is enumerate a few options and discuss their relative merits.

Let us consider interfaces that present both the matching documents and the faceted refinements in the same view. There are two conventional layouts: placing the facets in a panel to the left of the results and placing the facets directly above the results. An alternative approach is to place the facets directly below the results, but this approach has the drawback that users may not even be aware of the facets unless they scroll to the bottom of the page if the list of results pushes the faceted section below the fold.

Placing the results in the center and the facets on the left-hand side (as in Figure 7.1) makes it more likely that users will see search results and thus focus on them.

Users may prefer a view that brings them to the documents as quickly as possible, particularly if they only perform faceted refinement infrequently. For sophisticated users, this vertical layout

FIGURE 7.1: Results for "digital cameras" at www.jr.com.

makes both the results and facets immediately visible. However, this layout may be too subtle for less sophisticated users to notice the availability of the faceted refinement options. Also, such users may be unfamiliar with faceted refinement and may confuse the faceted refinement links with static site navigation links that lose their current query context.

Placing the facets above the results (as in Figure 7.2) makes it easy for users to notice them: they stand between the search box and the results, and they are above the fold. The downside is that

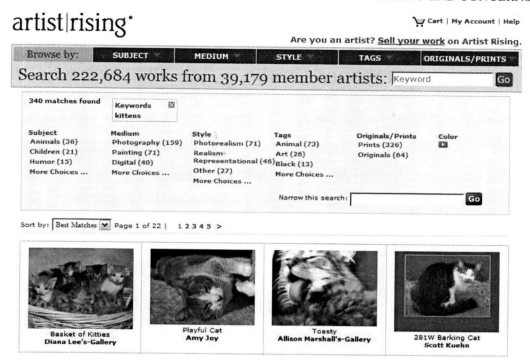

FIGURE 7.2: Results for "kittens" at www.artistrising.com.

placing the facets in such prime real estate means that users may need to make more effort to see the results. However, this effect may have a useful consequence: users may use a faceted search system more productively if they consider ways to elaborate their queries before rushing to the result list for their initial queries—and this strategy promotes the importance of the facets.

Another, less common, option is to present results and facets in separate views, for example, each in its own tab. The main choice in implementing such a design is to force which particular view the user sees first—similar to determining the layout for a view with results and facets presented on the same page. When users cannot see the results from their search, they are likely to make an effort to find them—unless they are so annoyed by the extra work they have to do that they abandon the site. Conversely, they are less likely to make such an effort to find facets that they may not even know exist.

Yet another possibility is to make the presentation of results facets a function of the query itself—that is, change the presentation based on the properties of the search results. Faceted refinement can accomplish at least one of two goals: clarification of ambiguous queries and refinement for general ones [102]. For ambiguous queries, it may be appropriate to present the user with a clarification dialog as soon as possible. For example, the search query "intelligence" on a library site might retrieve items about mental ability or information gathering; until the system has established the user's intended meaning, presenting results is a guessing game. For unambiguous queries, there is less urgency around offering a refinement dialog because the results will at least make sense to the user, even if they are not optimally relevant.

7.2   ORGANIZING FACETS AND FACET VALUES

In the previous chapter, we considered the problem of information overload from a back-end perspective: how do we prune the number of facets or facet values that we present to users?

In this section, we consider the corresponding front-end problem: how do we organize the information that we do present?

We approach the challenge of information overload in two steps: reducing the number of facets and values presented (as discussed in Section 6.3) and then organizing them as effectively as possible.

There are three general strategies for organizing facets, and these are similar to the strategies discussed in the previous chapter:

- Use a static order that does not change as the user navigates.
- Dynamically rank the order of presentation of facets based on their estimated utility to the user.
- Organize similar or related facets into groups.

The choice of using a static facet order vs. ranking facets is an interesting trade-off.

On one hand, a static order has the advantage or reinforcing the user's mental model—because the user will always see the same facets in the same order. This strategy works best when the number of facets is small in that all facets are visible at all times.

On the other hand, a static ordering is less effective when the number of facets is too large to be displayed at once or when some facets only apply to particular query contexts. For example, on an e-commerce site that sells consumer electronics, some facets only apply to small subsets of the product catalog, such as wattage for audio speakers or megapixels for digital cameras. Those facets should only be displayed when the user is looking at narrow result sets. In general, the same kinds of utility measures used for filtering can also be used for ranking.

Grouping related facets, as in ACM Digital Library's grouping related people, publications, and conferences (Figure 7.3), makes it possible to include more facets in a single display using less space on the page because the groups can be expanded and collapsed as desired by the user. Rich internet applications, designed using such technologies as AJAX (Asynchronous JavaScript and

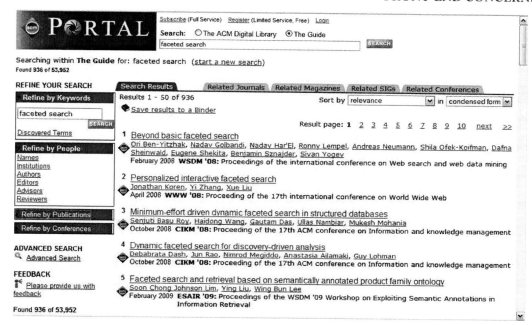

FIGURE 7.3: Results for "faceted search" in the ACM Digital Library.

XML) or Flash, make it easier to implement such an approach that still offers users a highly responsive interface.

Similar strategies apply for presenting the facet values:

- Use a static order that is independent of the query context.
- Rank facets based on a utility measure.
- Present hierarchical facet values progressively, that is, one level of the hierarchy at a time.

This last strategy is an option even when the facet is not hierarchical. We can create an artificial hierarchy in which no node has more than a tractable number of children, for example, 5 or 10.

For example, consider a facet with a large set of strings as values, such as book authors. We can divide up them up by last name, splitting them at the top level as A–E, F–K, L–P, Q–U, and V–Z. The A–E node can be split into A, B, C, D, and E; A split into Aa–Ae, Af–Ak, etc.; and so forth until no node has more than five authors as (leaf) children. The design of such an artificial hierarchy requires some care to mitigate the trade-off between excessive depth and excessive fan-out at each node.

We can take a similar approach to create a hierarchy of ranges for numerical values or for facets whose values can be clustered based on a similarity measure.

7.3   THE SEARCH BOX

Although our discussion of faceted search has focused on the use of facets for refinements, it is important to remember that faceted search is still a type of search—and that often the entry point into a faceted search system is the search box. Without the search box, we would only have a faceted navigation system—a useful interface for many applications but too limited to enjoy the broad success of faceted search.

Combining free-text search and faceted refinement is powerful: it allows users to create semi-structured queries and thus access structured and unstructured content. However, the search box also raises significant design challenges for application developers.

The designer of a faceted search system must make a number of choices about how the search box behaves:

- Should a search query adhere to the current query filters?
- Should search look at all of the text in each document, or should search be restricted to specific fields?
- How should the search handle multiword queries by default? Should the words be combined as an OR (i.e., match any word), as an AND (i.e., match all words), or as a phrase (i.e., the words must all occur in a document and in that exact sequence), and should the user be given a choice in this matter?
- Should search queries be subject to query expansion, such as matching words that are variants of query terms?
- Should systems present multiple search boxes, a parameterizable search box, or an advanced search interface?

These are open-ended questions and only represent a subset of the questions about search behavior that face designers of faceted search applications. We will try to supply some answers—or, at least, guidance.

First, let us consider the question of whether the search query should respect the current query filters. In the most common use case for faceted search, a user initially enters a free-text search query and then follows it up by one or more refinements using the facets. For example, a user types in "digital cameras" and refines on a specific megapixel range. But how do we handle deviations from this common case, for example, when the user first narrows the document collection by selecting a facet value and then performs a free-text search? Does the text search adhere to the faceted refinement or start a new query from scratch?

The conventional and probably safest approach is for free-text search to default to clearing all other filters or to offer users the options to search within the current results, for example, by clicking on a check box indicating that the user is explicitly choosing to search only within the set of results that is currently being viewed. The interface for the Triangle Research Libraries Network in Section 5.1 illustrates this check-box approach.

Figure 7.4 suggests an alternate approach, preserving query context by default and signaling this behavior to users by labeling the search box with "search within these results" before a user starts typing.

Now let us consider the question of whether we should match a search query against the full document text or only a restricted set of text fields. A common approach for search engines is to perform search against full text and to rely on relevance ranking to push more relevant results to the top of the results.

Although this approach may work well in a conventional search engine, it can undermine the effectiveness of faceted search. As we discussed in Chapter 3, faceted search builds on a set retrieval model: the faceted refinements reflect all of the results not just the results that the system judges to be most relevant. If most of the results are not relevant, then the faceted refinements may not be especially useful, especially if they are presented with counts indicating their distribution over the result set.

An alternative approach is for the default behavior to err on the side of precision, for example, searching only against the title field unless the user explicitly asks to include other fields. The key

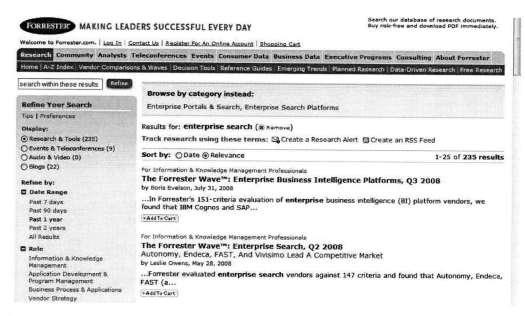

FIGURE 7.4: Search within results at www.forrester.com.

design question is whether the benefit of increased precision cost usually outweighs the cost of decreased recall. It is also possible to hedge, to obtain search results from a broader search query that favors recall, and to derive the faceted refinements from a narrower search query that favors precision.

The search box can also serve as a way for the user to search the set of facets, not just the documents themselves, as shown in Figure 7.5 (the "categories" are facet values). An advantage of this approach is that the set of documents assigned a particular facet value is often a more accurate result set than the set of documents containing those words in their text. It is even possible to search against the set of combinations of facet values [103].

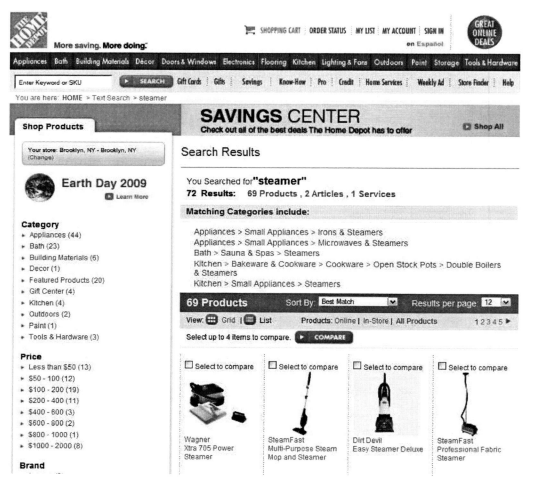

FIGURE 7.5: Results for "steamer" at www.homedepot.com.

Multiword search queries and query expansion also raise issues of precision and recall. However, because of their familiarity with web search engines, most users have become accustomed to search engines that interpret multiword search queries as an unordered conjunction (an AND, not an OR), for example, a search for *faceted navigation* returns documents containing both words but not necessarily in that order or even next to each other. While we should not be fatalistic about conventions, we must recognize that flouting them will incur some amount of user confusion.

The conventions for query expansion are less established, but most users expect, at a minimum, that they do not need to worry whether they use the singular or plural form of a noun (i.e., that both will return the same results). More aggressive query expansion (e.g., employing a thesaurus to obtain additional matches for words related to the query terms) is again a precision/recall trade-off. More importantly, it calls for transparency to avoid confusing users with unexpected and unexplained results. Any expansion that is unintuitive to a user is not worth the risk of confusing users and thus undermining their faith in the system.

Finally, there is the question of whether to offer users multiple search boxes, a parameterizable search box, or an advanced search interface. There are no hard and fast rules here, but multiple search boxes with different behaviors have the potential to confuse users. A few users will configure a parameterizable search box, but most will never change the default search behavior. Hence, it is critical that the default search behavior be reasonable. Similarly, whereas a minority of users will appreciate the opportunity to use an advanced (typically parametric) search interface, many will never even discover it exists. To avoid confusing the majority of users, many successful retrieval applications place their advanced search interface on a separate page.

7.4 MULTIPLE SELECTION FROM A FACET

The most common use case for faceted search or navigation is to select at most one value per facet, but there are at least two ways from which a user might select multiple values from the same facet:

- Disjunctive (OR) selection. Selecting a range (e.g., a price or date range) may be a kind of disjunctive selection, depending how the values are represented.
- Conjunctive (AND) selection.

The design challenge is to communicate to users whether selecting multiple values from a particular facet is disjunctive or conjunctive—particularly if the site offers both behaviors. Users are notoriously bad at inferring Boolean logic from subtle cues.

It is important to use an interface that not only is self-consistent but also adheres to familiar conventions. For example, the check boxes in Figure 7.6 adheres to the convention for disjunctive selection.

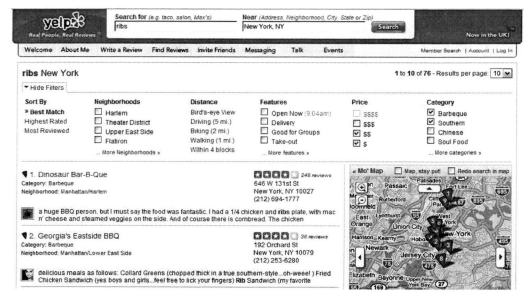

FIGURE 7.6: Results for "ribs" in New York at www.yelp.com.

There are fewer interfaces that allow conjunctive selection from the same facet, but a convention for those that do is to present the selections as ordinary links, as with the topic facet in Figure 7.7.

The approach may make the user think that he or she is drilling down a hierarchy, but fortunately that misinterpretation is consistent with the narrowing effect of conjunctive selection.

Perhaps most importantly, we urge caution in combing disjunctive and conjunctive selection in the same interface. Users who can understand such a complex process will be better served by the ability to construct Boolean queries at a command line.

A rule of thumb is that facets that are typically singly assigned to documents (e.g., brand, document type) work well with disjunctive selection, whereas facets that are often multiply assigned to documents (e.g., consumer electronics features, topic) work well with conjunctive selection.

7.5 DESIGN PATTERNS

The previous sections have focused on specific aspects of faceted search interfaces that raise front-end usability concerns. A more holistic approach to these concerns is to follow design patterns.

A concept that originated in software engineering, a *design pattern* is a general, reusable solution to a commonly occurring design problem. The definitive textbook on design patterns is a book

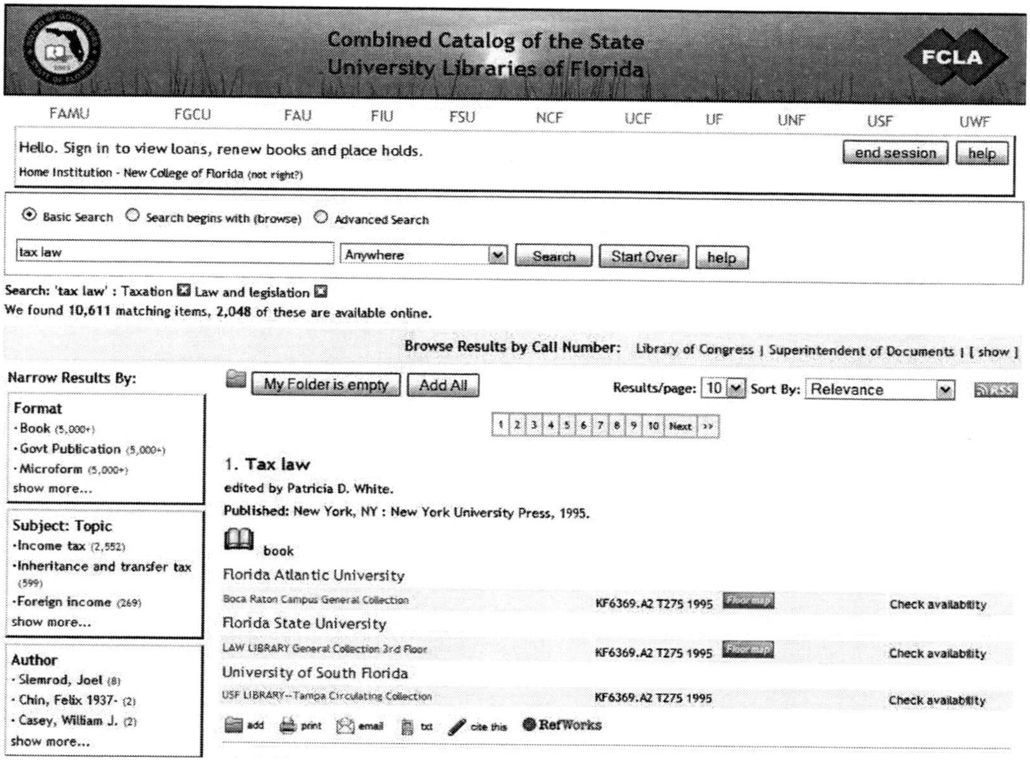

FIGURE 7.7: Results for "tax law" at www.fcla.edu.

of that title, written by the illustrious "Gang of Four" [104]. Today, the term *architectural pattern* refers more to a generally established solution for a problem in software architecture.

Although faceted search is relatively new, particularly in terms of mainstream adoption, there are already collections of design patterns for it. In particular, Peter Morville maintains a collection of search patterns that address faceted search [105].

Other designers who have built pattern libraries include Martijn Van Welie [106] and Janne Lammi [107]. The Yahoo! Design Pattern Library [108] more generally addresses the design of Web sites, but some of its patterns are useful for faceted search applications in particular.

Front-end design is a complicated, open-ended challenge, and the best we can hope to offer are rough guidelines. Pattern libraries assemble best practices from deployed applications and thus complement theoretically motivated advice from the wisdom of practical experience. Moreover, because they use concrete examples, they may supply details that, although minor, make crucial differences to the look and feel of a faceted search site. We encourage readers to learn from and contribute to pattern libraries.

7.6   TAKE-AWAYS

- The power of faceted search can overwhelm and confuse users if it is implemented with a poor design.
- Choosing the correct layout of facets and results in an interface can be a trade-off between the work required for users to see results and the likelihood they notice the facets.
- For ambiguous queries, users may benefit from a facet-driven clarification dialog.
- Strategies to avoid information overload by filtering facets and facet values also offer ways to rank and organize them.
- Users generally expect that initiating a new free-text search will clear current query filters, but some applications provide the option to search within current results.
- A number of decisions about search behavior involve a precision/recall trade-off: consider computing results to favor recall but computing the utility of facets and facet values based on a narrower search query that favors precision.
- An interface allowing multiple selections within a single facet should not only be self-consistent but adhere to familiar conventions.
- Consider the use of design patterns to take a holistic approach and learn from the collective wisdom of practitioners.

CHAPTER 8

Conclusion

"A conclusion is the place where you got tired of thinking."

—Arthur Bloch

Congratulations, you have made it to the end of the lecture! I hope it's been an enlightening journey from Aristotle's tree to Ranganathan's colons—and then all the way to modern web sites and the practical considerations for implementing a faceted search system that delivers a satisfying user experience. As someone who has spent the past decade working on faceted search, I cannot help but be an evangelist for an approach that, in my view, is critical to building support for the rich spectrum of information-seeking needs.

Where do we go from here? I hope this lecture has raised at least as many questions as it has answered. Like most important ideas, faceted search takes a day to learn but a lifetime to master. Well, perhaps not a lifetime—it is hard to make such a claim about an idea that has only been around for a generation.

Faceted search is a very young research topic. Only now are we seeing formal analysis of faceted search interface using the rigor applied to other query models, such as those for databases and data warehouses [109]. Also, compared to the mature evaluation methodology for information retrieval systems, evaluation of faceted search is still nascent [110].

A particularly interesting question is whether we will ever see faceted search applied to the open Web. There are major challenges to surmount [111], but exploratory search engines like Kosmix [112] and Cuil [113] are at least taking the first steps toward surmounting these challenges.

Faceted search and exploratory search are part of a broader program of human–computer information retrieval (HCIR) that applies interactive techniques to a broad spectrum of information-seeking tasks. Although the best-first search approach of modern search engines has addressed many of our needs, this approach has reached the point of diminishing returns. HCIR embraces interaction as the foundation for information-seeking support, and faceted search is a key enabler for such interaction.

Some additional resources that may prove useful are the proceedings of recent workshops, all available online:

- HCIL 2005: Exploratory Search Interfaces [114]
- SIGIR 2006: Faceted Search [115]
- SIGCHI 2007: Exploratory Search and HCI [116]
- HCIR 2007 and 2008 [117, 118]
- Information Seeking Support Systems 2008 [119]

I hope this lecture has persuaded you of the merits of the HCIR program in general, and of faceted search in particular. Perhaps it will even inspire you to help advance that program by building out the necessary models, evaluation frameworks, and tools to support it. It has been my pleasure to take you on this journey, and I hope you have enjoyed reading it as much as I have enjoyed writing it.

*　*　*　*

Glossary

Boolean retrieval   Set retrieval based on Boolean expressions using operators such as AND, OR, and NOT.

conjunctive selection   "AND" or "match all" selection for mutliple-word search queries.

design pattern   General, reusable solution to a commonly occurring design problem.

disjunctive selection   "OR" or "match any" selection for mutliple-word search queries.

exploratory search   Information seeking where the user does not have a known target document.

facet   Composable classification elements that can be used to construct compound subjects.

faceted navigation   Interface allows user to elaborate a faceted query progressively.

faceted search   Interface that combines faceted navigation with free-text search.

metadata   Structured attributes associated with a text document.

ontology   Representation of a set of concepts and the relationships among those concepts.

parametric search   Interface for faceted content allowing users to query by specifying constraints on facet values.

precision   Fraction of retrieved documents that are relevant to an information need.

query expansion   Expanding a search query to include additional words or phrases related to the query.

ranked retrieval   Document retrieval that returns results ordered by estimated relevance.

recall   Fraction of all possible relevant documents to an information need that are retrieved.

relevance   Measure of information conveyed by a document relative to user's information need.

set retrieval   Document retrieval that returns results as unordered document sets.

taxonomy   Hierarchical classification scheme for representing knowledge.

vocabulary problem   Keywords assigned by indexers are often different than those used by searchers.

References

[1] Bush, V. 1945 (July). As we may think. *The Atlantic Monthly 176(1)*: pp. 101–108.

[2] Reitz, J. 2004. *Dictionary for Library and Information Science*. Westport, CT: Libraries Unlimited.

[3] Bush, V. 1945. As we may think: a top U.S. scientist foresees a possible future world in which man-made machines will start to think. *Life 19(11)*: pp. 112–124.

[4] White, R., and Roth, R. 2009. *Exploratory Search: Beyond the Query-Response Paradigm*. San Rafael, CA: Morgan and Claypool.

[5] Marchionini, G. 2006 (June/July). Toward human–computer information retrieval. *Bulletin of the American Society for Information Science*. www.asis.org/Bulletin/Jun-06/marchionini.html/.

[6] English, J., Hearst, M., Sinha, R., Swearingen, K., and Yee, P. 2002. Flexible search and navigation using faceted metadata. Unpublished data.

[7] Porter, J. 2003. Testing the three-click rule. *User Interface Engineering*. http://www.uie.com/.

[8] Neill, A., and Ridley, A. 1995. *The Philosophy of Art: Readings Ancient and Modern*. New York: McGraw-Hill.

[9] Whitehead, A. 1929. *Process and Reality*. New York: Macmillan.

[10] http://www.lmpc.edu.au/resources/science/livingthings/.

[11] Aristotle. 343 BC. *Historia Animalium*.

[12] Aristotle. 350 BC. *De Partibus Animalium*.

[13] Aristotle. 350 BC. *De Generatione Animalium*.

[14] Linnaeus, C. 1767. *Systema Naturae*.

[15] Chambers, E. 1728. *Cyclopaedia, or, An universal dictionary of arts and sciences*. London: James and John Knapton.

[16] Soergel, D. 1975. Theoretical problems of thesaurus building with particular reference to concept formation. In Petoefi et al. (Ed.), *Fachsprache-Umgangssprache*. Kronberg, Ts: Scriptor; 1975.

[17] Dewey, M. 1876. *A Classification and Subject Index for Cataloguing and Arranging the Books and Pamphlets of a Library*. Amherst, MA: Case, Lockwood & Brainard.

[18] http://dir.yahoo.com/.

[19] http://www.dmoz.org/.

[20] *Dewey Decimal Classification and Relative Index, Edition 22 (DDC 22).* Dublin, OH: OCLC Forest Press.

[21] Parkhi, R. S. 1964. *Decimal Classification and Colon Classification in Perspective.* London: Asia Publishing House, p. 130.

[22] Ranganathan, S. R. 1951. *Philosophy of Library Classification.* Copenhagen: Ejnar Munksgaard.

[23] Cutter, C. 1891–1893. *Expansive Classification, Part 1: The First Six Classifications.* Boston, MA: Cutter.

[24] Ranganathan, S. R. 1933. *Colon Classification.* Madras, India: Madras Library Association.

[25] Ranganathan, S. R. 1950. *Classification, Coding, and Machinery for Search.* Paris: UNESCO.

[26] http://dublincore.org/.

[27] Gruber, T. 1993. A translation approach to portable ontology specifications. *Knowledge Acquisition 5*: pp. 199–220.

[28] Smith, M., Welty, C., and McGuinness, D. L. 2004. OWL Web Ontology Language Guide. W3C.

[29] Herman, I. 2008. Semantic Web Activity Statement. W3C.

[30] Singhal, A. 2001. Modern information retrieval: a brief overview. *Bulletin of the IEEE Computer Society Technical Committee on Data Engineering 24(4)*: pp. 35–43.

[31] Goffman, W. 1964. On relevance as a measure. *Information Storage and Retrieval 2(3)*: pp. 201–203.

[32] Mizzaro, S. 1997 (September). Relevance: the whole history. *Journal of the American Society for Information Science 48(9)*: pp. 810–832.

[33] Saracevic, T. 2006. Relevance: a review of the literature and a framework for thinking on the notion in information science. Part II. *Advances in Librarianship 30*: pp. 3–71.

[34] http://trec.nist.gov/.

[35] Voorhees, E. 2008 (November). On test collections for adaptive information retrieval. *Information Processing and Management 44(6)*: pp. 1879–1885.

[36] Manning, C., Raghavan, P., and Schütze, H. 2008. *Introduction to Information Retrieval.* New York: Cambridge University Press.

[37] http://patft.uspto.gov/netahtml/PTO/search-adv.htm.

[38] Spink, A., Wolfram, D., Jansen, M. B., and Saracevic, T. 2001. Searching the Web: the public and their queries. *Journal of the American Society for Information Science and Technology 52,3*: 226-234.

[39] Robertson, S. 2000 (April). Salton Award Lecture on theoretical argument in information retrieval. *SIGIR Forum 34(1)*: pp. 1–10.

[40] Turtle, H. 1994. Natural language vs. Boolean query evaluation: a comparison of retrieval performance. In *Proceedings of the 17th Annual international ACM SIGIR Conference on Research and Development in information Retrieval*, pp. 212–220.

[41] Mooers, C. 1950. Coding, information retrieval, and the rapid selector. *American Documentation 1(4)*: pp. 225–229.

[42] Mooers, C. 1961. From a point of view of mathematical etc. techniques. In R. A. Fairthorne (Ed.), *Towards Information Retrieval*, pp. xvii–xxiii. London: Butterworths.

[43] Salton, G., Wong, A., and Yang, C. S. 1975. A Vector Space Model for Automatic Indexing. *Communications of the ACM 18(11)*: pp. 613–620.

[44] Spärck Jones, K. 1972. A statistical interpretation of term specificity and its application in retrieval. *Journal of Documentation 28(1)*: pp. 11–21.

[45] Deerwester, S., Dumais, S., Furnas, G. W., Landauer, T. K., and Harshman, R. 1990. Indexing by latent semantic analysis. *Journal of the American Society for Information Science 41(6)*: pp. 391–407.

[46] Kleinberg, J. 1999. Authoritative sources in a hyperlinked environment. *Journal of the ACM (JACM) 46(5)*: pp. 604–632.

[47] Brin, S., and Page, L. 1998. The anatomy of a large-scale hypertextual Web search engine. *Computer Networks and ISDN Systems30(1–7)*: pp. 107–117.

[48] The History of Yahoo!—How It All Started. http://docs.yahoo.com/info/misc/history.html.

[49] Furnas, G., Landauer, T., Gomez, L., and Dumais, S. 1987. The Vocabulary Problem in Human-System Communication. *Communications of the ACM 30(11)*: pp. 964–971.

[50] Perugini, S. 2008. Symbolic links in the Open Directory Project. *Information Processing and Management 44(2)*: pp. 910–930.

[51] Neal, M. 1997. Parametric search: evolving information retrieval for the web. http://www.intranetjournal.com/ features/cadis-1.shtml.

[52] Permission obtained from Endeca Technologies, Inc. http://endeca.com/.

[53] Shneiderman, B. 1994 (November). Dynamic queries for visual information seeking. *IEEE Software 11(6)*: pp. 70–77.

[54] Date, C. J. 1989. *A Guide to the SQL Standard*. Boston, MA: Addison-Wesley Longman Publishing Co., Inc.

[55] Ahlberg, C., and Shneiderman, B. 1994. Visual information seeking: tight coupling of dynamic query filters with starfield displays. In *Proceedings of the SIGCHI Conference on Human Factors in Computing Systems: Celebrating Interdependence* (Boston, MA, USA, April 24–28, 1994). B. Adelson, S. Dumais, and J. Olson (Eds.), CHI '94. pp. 313–317. New York: ACM.

[56] Doan, K., Plaisant, C., and Shneiderman, B. 1996. Query previews in networked information systems. In *Proceedings of the 3rd International Forum on Research and Technology Advances in Digital Libraries (May 13–15, 1996)*, ADL. p. 120. Washington, DC: IEEE Computer Society.

[57]   Pollitt, A. S., Smith, M., Treglown, M., and Braekevelt, P. 1996. View-based searching systems—progress towards effective disintermediation. *Online Information 96 Proceedings.* pp. 433–441.

[58]   Hearst, M., Pedersen, J., and Karger, D. 1995. Scatter/Gather as a tool for the analysis of retrieval results. *Working Notes of the AAAI Fall Symposium on AI Applications in Knowledge Navigation.*

[59]   Hearst, M. 2000. Next generation web search: setting our sites. *IEEE Data Engineering Bulletin 23(3):* pp. 38–48.

[60]   http://sourceforge.net/projects/flamenco.

[61]   Hearst, M. 2006. Design recommendations for hierarchical faceted search interfaces. *ACM SIGIR Workshop on Faceted Search.*

[62]   Stoica, E., and Hearst, M. 2004. Nearly-automated metadata hierarchy creation. In *Companion Proceedings of HLT-NAACL'04.*

[63]   Stoica, E., Hearst, M., and Richardson, M. 2007. Automating creation of hierarchical faceted metadata structures. In *Proceedings of NAACL-HLT.*

[64]   Hearst, M. 2006. Clustering versus faceted categories for information exploration. *Communications of the ACM, 49(4):* pp. 56- 61.

[65]   Capra, R., and Marchionini, G. 2008. The Relation Browser Tool for Faceted Exploratory Search. In *Proceedings of the 2008 Conference on Digital Libraries (JCDL '08)*, Pittsburgh, Pennsylvania, June 16–20, 2008.

[66]   Marchionini, G., and Brunk, B. (2003). Towards a general relation browser: a GUI for information architects. *Journal of Digital Information, 4(1).*

[67]   Zhang, J., & Marchionini, G. (2004). Relational Browser++: a fast and contextualized searching and browsing tool. Technical Report, SILS-TR-2004-01.

[68]   Schraefel, M. C., Karam, M., and Zhao, S. 2003. mSpace: interaction design for user-determined, adaptable domain exploration in hypermedia. In: *AH 2003: Workshop on Adaptive Hypermedia and Adaptive Web Based Systems*, August 26, Nottingham, UK.

[69]   Schraefel, M. C., Smith, D. A., Russel, A., Owens, A., Harris, C., and Wilson, M. L. 2005. The mSpace classical music explorer: improving access to classical music for real people. In *V MUSICNETWORK OPEN WORKSHOP: Integration of Music in Multimedia Applications,* July 4 and 5, 2005, Vienna, Austria.

[70]   Huynh, D., and Karger, D. 2009. Parallax and Companion: set-based browsing for the data web. Submitted to *WWW 2009.*

[71]   http://www.freebase.com/.

[72]   Anderson, C. 2007. Record relationship navigation: implications for information access and discovery. *Workshop on Human–Computer Interaction and Information Retrieval (HCIR 2007).*

[73] http://www.endeca.com/.

[74] http://search.trln.edu/.

[75] http://www.lib.ncsu.edu/endeca/.

[76] http://www.trln.org/highlights/feb2007.pdf.

[77] http://www.ebay.com/.

[78] Ebay's Express to take on Amazon. 2006. *The Sunday Times.* October 1, 2006.

[79] Too Smart for its Own Good. 2008. *Forbes.* October 2, 2008.

[80] Hearst, M. 2006. Visualization in text analysis problems. *VAC Consortium Meeting.*

[81] http://www.internetretailer.com/top500/list.asp.

[82] Cox, B. 2002. It's official, it's called ruby, it's in beta. *ecommerce-guide.com.* November 1, 2002.

[83] http://wiki.apache.org/solr/.

[84] Hostetter, C. 2006. Faceted searching with Apache Solr. *ApacheCon US 2006.*

[85] http://drupal.org/.

[86] Strohman, T. 2007. Efficient processing of complex features for information retrieval. Doctoral dissertation, Amherst, MA: University of Massachusetts.

[87] http://yonik.wordpress.com/2008/11/25/solr-faceted-search-performance-improvements/.

[88] Hennessy, J. L., and Patterson, D. A. 1996. *Computer Architecture (2nd Ed.): A Quantitative Approach.* Boston, MA: Morgan Kaufmann.

[89] Knabe, F., and Tunkelang, D. 2004. Processing search queries in a distributed environment. In *Proceedings of the Thirteenth ACM international Conference on information and Knowledge Management* (Washington, DC, USA, November 08–13, 2004). *CIKM '04.* pp. 492–494. New York: ACM.

[90] Simon, H. 1971. Designing organizations for an information-rich world. In *Computers, Communication, and the Public Interest.* Baltimore, MD: The Johns Hopkins University Press.

[91] Cubranic, D. 2008. Polestar: assisted navigation for exploring multi-dimensional information spaces. *Workshop on Human-Computer Interaction in Information Retrieval (HCIR'08).*

[92] Koren, J., Leung, A., Zhang, Y., Maltzahn, C., Ames, S., and Miller, E. 2007. Searching and navigating petabyte-scale file systems based on facets. *Proceedings of the 2nd international Workshop on Petascale Data Storage (PDSW '07):* pp. 21–25.

[93] Smith, G. 2007 *Tagging: People-Powered Metadata for the Social Web.* Berkeley, CA: New Riders Publishing.

[94] Weiss, S., Indurkhya, N., Zhang, T., and Damerau, F. 2004. *Text Mining: Predictive Methods for Analyzing Unstructured Information.* New York: Springer-Verlag.

[95] http://www.cs.technion.ac.il/~gabr/resources/atc/atcbib. Html.

[96] https://www.mturk.com/.

[97]   Zelevinsky, V., Wang, J., and Tunkelang, D. 2008. Supporting exploratory search for the ACM Digital Library. *Workshop on Human–Computer Interaction and Information Retrieval (HCIR'08)*, October 2008.

[98]   Von Ahn, L. 2006. Games with a purpose. *Computer 29(6)*: pp. 92–94.

[99]   Furnas, G. 1985. Experience with an adaptive indexing scheme. *Human Factors in Computing Systems CHI '85 Conference Proceedings*. pp. 131–135.

[100] Baeza-Yates, R. A., and Ribeiro-Neto, B. 1999. *Modern Information Retrieval*. Boston, MA: Addison-Wesley Longman Publishing Co., Inc.

[101] Pirolli, P., Card, S., and Van Der Wege, M. 2000 (May). The effect of information scent on searching information: visualizations of large tree structures. *Advanced Visual Interfaces*.

[102] http://thenoisychannel.com/2008/06/02/clarification-vs-refinement/.

[103] Tunkelang, D. 2006. Dynamic category sets: an approach for faceted search. In *SIGIR 2006 Faceted Search Workshop*.

[104] Gamma, E., Helm, R., Johnson, R., and Vlissides, J. 1995. *Design Patterns: Elements of Reusable Object-Oriented Software*. Boston, MA: Addison-Wesley Longman Publishing Co., Inc.

[105] http://www.flickr.com/photos/morville/collections/72157603785835882/.

[106] Stefaner, M., and Muller, B. 2007. Elastic lists for facet browsers. *Proceedings of the 18th international Conference on Database and Expert Systems Applications*. pp. 217–222.

[107] http://uipatternfactory.com/.

[108] http://developer.yahoo.com/ypatterns/.

[109] Clarkson, E. 2009. Generalized formal models for faceted user interfaces. Accepted for publication in *Proceedings of JCDL '09*. JCDL 2009 is June 15–19, 2009 (http://www.jcdl2009.org/).

[110] Wilson, M., and schraefel, m. 2007. Bridging the gap: using IR models for evaluating exploratory search interfaces. *SIGCHI 2007 Workshop on Exploratory Search and HCI*. San Jose, CA.

[111] Teevan, J., Dumais, S., and Gutt, Z. 2008. Challenges for supporting faceted search in large, heterogeneous corpora like the Web. In *Proceedings of HCIR 2008*.

[112] http://kosmix.com/.

[113] http://cuil.com/.

[114] http://research.microsoft.com/en-us/um/people/ryenw/xsi/.

[115] http://facetedsearch.googlepages.com/.

[116] http://research.microsoft.com/en-us/um/people/ryenw/esi/.

[117] http://projects.csail.mit.edu/hcir/.

[118] http://research.microsoft.com/en-us/um/people/ryenw/ hcir2008/.

[119] http://www.ils.unc.edu/ISSS_workshop/.

Author Biography

Daniel Tunkelang is co-founder and chief scientist at Endeca, a leading provider of enterprise information access solutions. He heads up Endeca's efforts to develop features and capabilities that emphasize interactive information retrieval.

Daniel is recognized as a leading advocate of human–computer information retrieval, a multidisciplinary effort to bridge the gap between the more systems-oriented work in information retrieval and the more cognitively focused approach in library and information science. He has organized annual workshops on the subject. He publishes *The Noisy Channel*, a widely read and cited blog on the information-seeking process. He also participates actively in both academic and industry conferences, recently attempting to bridge the gap between the two by organizing an Industry Track at SIGIR, the leading academic conference on information retrieval.

Daniel holds undergraduate degrees in mathematics and computer science from MIT and a PhD in computer science from CMU. Before co-founding Endeca, he worked at the IBM T. J. Watson Research Center and AT&T Labs. He lives in New York with his wife Kristin and daughter Lily.